MATHEMATICAL CHALLENGES FROM THEORETICAL/COMPUTATIONAL CHEMISTRY

Committee on Mathematical Challenges from Computational Chemistry

Board on Mathematical Sciences
and
Board on Chemical Sciences and Technology

Commission on Physical Sciences, Mathematics, and Applications

National Research Council

National Academy Press
Washington, D.C. 1995

NOTICE: The project that is the subject of this report was approved by the Governing Board of the National Research Council, whose members are drawn from the councils of the National Academy of Sciences, the National Academy of Engineering, and the Institute of Medicine. The members of the committee responsible for the report were chosen for their special competences and with regard for appropriate balance.

This report has been reviewed by a group other than the authors according to procedures approved by a Report Review Committee consisting of members of the National Academy of Sciences, the National Academy of Engineering, and the Institute of Medicine.

Support for this project was provided by the National Science Foundation, Air Force Office of Scientific Research, Department of Energy, Army Research Office, and Ciba-Geigy Corporation.

Library of Congress Catalog Card Number 95-68040
International Standard Book Number 0-309-05097-9

Available from:
National Academy Press
2101 Constitution Avenue, NW
Box 285
Washington, DC 20055
B-515
Available on the Internet via the World Wide Web at the URL: http://www.nas.edu/.

Printed in the United States of America

COVER ILLUSTRATION: The catalytic binding site of the enzyme purine nucleoside phosphorylase, which plays a key role in immune function, is shown in gray as a space-filling model. An inhibitor of the enzyme is shown in white. A tight fit between the enzyme and the inhibitor is required for binding and inhibitory activity, and a goal of structure-based drug discovery is the design of inhibitors that are geometrically (and chemically) complementary to an enzyme binding site.

The figure was computer generated and resulted from a study that involved calculating geometries of potential inhibitors "docked" in the enzyme binding site (Montgomery et al., 1993). The study involved energy minimization and Monte Carlo-like conformational searching using the MacroModel computational chemistry software (Mohamadi et al., 1990). Such a computationally intensive task could not have been carried out 10 years ago and was an integral part of a structure-based drug design effort (Montgomery, 1993; see also Bugg et al., 1993).

Figure courtesy of W. Guida, Pharmaceuticals Division, Ciba-Geigy Corporation.

References

Bugg, C.E., W.M. Carson, and J.A. Montgomery, 1993, Drugs by design, *Scientific American* 269:92–98.

Mohamadi, F., N.G.J. Richards, W.C. Guida, R. Liskamp, M. Lipton, C. Caufield, G. Chang, T. Hendrickson, and W.C. Still, 1990, MacroModel—An integrated software system for modeling organic and bioorganic molecules using molecular mechanics, *J. Comput. Chem.* 11:440–467.

Montgomery, J.A., 1993, Purine nucleoside phosphorylase: A target for drug design, *Medicinal Research Reviews* 13:209–228.

Montgomery, J.A., S. Niwas, J.D. Rose, J.A. Secrist, Y.S. Babu, C.E. Bugg, M.D. Erion, W.C. Guida, and S.E. Ealick, 1993, Structure-based design of inhibitors of purine nucleoside phosphorylase 1. 9-(arylmethyl) derivatives of 9-deazaguanine, *J. Med. Chem.* 36:55–69.

Committee on Mathematical Challenges from Computational Chemistry

FRANK H. STILLINGER, AT&T Bell Laboratories, *Chair*
HANS C. ANDERSEN, Stanford University
LOUIS AUSLANDER, City University of New York
DAVID L. BEVERIDGE, Wesleyan University
ERNEST R. DAVIDSON, Indiana University
WAYNE C. GUIDA, Ciba-Geigy Corporation
PETER A. KOLLMAN, University of California at San Francisco
WILLIAM A. LESTER, JR., University of California at Berkeley
YVONNE C. MARTIN, Abbott Laboratories
GEORGE C. SCHATZ, Northwestern University
TAMAR SCHLICK, New York University and Howard Hughes Medical Institute
L. RIDGWAY SCOTT, University of Houston
DEWITT L. SUMNERS, Florida State University
PETER G. WOLYNES, University of Illinois at Urbana-Champaign

Board on Chemical Sciences and Technology Liaison
KENDALL N. HOUK, University of California at Los Angeles

SCOTT T. WEIDMAN, Study Director
TAÑA L. SPENCER, Project Assistant

Board on Chemical Sciences and Technology

Commission on Physical Sciences, Mathematics, and Applications

CONTENTS

LIST OF BOXES

EXECUTIVE SUMMARY

Overview

Although much of its discovery process is descriptive and qualitative, chemistry is fundamentally a quantitative science. It serves a wide range of human needs, activities, and concerns. The mathematical sciences provide the language for quantitative science, and this language is growing in many directions as computational science in general continues its rapid expansion. A timely opportunity now exists to strengthen and increase the beneficial impacts of chemistry by enhancing the interaction between chemistry and the mathematical sciences.

Computational chemistry is a natural outgrowth of theoretical chemistry, the traditional role of which involves the creation and dissemination of a penetrating conceptual infrastructure for the chemical sciences, particularly at the atomic and molecular levels. The mathematical sciences have been indispensable allies and have provided vital tools for that role. Theoretical chemistry has also sought to devise and to implement quantitative algorithms for organizing massive amounts of data from the laboratory, and for predicting the course and extent of chemical phenomena in situations that are difficult or even impossible to observe directly; thus, today it is difficult to classify many lines of research as either "theoretical" or "computational." This report tends toward the term theoretical/computational—any distinction between the two areas is rather misleading because the subject demands both quantitative characterization and conceptual understanding.

Computational chemistry has its roots in the early attempts by theoretical physicists, beginning in 1928, to solve the Schrödinger equation using hand-cranked calculating machines. By the 1950s, with the appearance of digital computers, serious attempts were being made to obtain highly accurate quantitative information about the chemical behavior of molecules via numerical approximations to the solution of the Schrödinger equation. In subsequent years, thanks to leaps in computing power and algorithms, methods have evolved from those that were used to study 1- and 2-atom systems in 1928, through those that were used to study 2- to 5-atom systems in 1970, to the present programs that produce useful quantitative results for molecules with up to 10 to 20 atoms. Other chemists, whose research can be accomplished with cruder models of the atom, have pushed this limit much higher. For instance, simpler approximations have long been used in the molecular mechanics approach that make possible the modeling of biological molecules with thousands of atoms.

Recent decades have witnessed a revolutionary expansion in the breadth and capability of theoretical and computational chemistry—with a commensurate rise in optimism regarding the ability of theoretical/computational chemistry to resolve pressing problems both of a fundamental scientific character and of clearly practical interest. Those outside the field may not realize that theoretical/computational chemistry, broadly defined, underpins rational drug design, contributes to the selection and synthesis of new compounds, and guides the design of catalysts. New quantum mechanical techniques underlie the understanding of electronic properties of materials and have advanced the level of precision at which molecules of at least moderate size can be modeled. Furthermore, computational chemistry software is a set of tools used increasingly by chemists of many persuasions. These various abilities and facilities have proved to be very important to U.S. industry, and their advancement would generate even further industrial benefits. Engaging problems and deep challenges for mathematical scientists are posed by the needs of theoretical and computational chemists, and the products of mathematical research in these areas could have far-reaching ramifications.

The statement of task given to the Committee on Mathematical Challenges from Computational Chemistry reads as follows: "The committee will investigate and report on opportunities for collaborative and synergistic research in the mathematical sciences that can accelerate progress in theoretical and computational chemistry and their applications, and make recommendations for promoting this research." It was clear from the outset that the study could not presume to be exhaustive. However, it seemed realistic to strive for representative sampling of the two communities involved and to identify instructive examples of past collaborative successes, likely prospects for interdisciplinary synergy, and barriers to joint research that could be removed or at least lowered.

A number of fruitful collaborations between mathematical scientists and theoretical/computational chemists have occurred in the past. Noteworthy examples include the Nobel prize-winning work of Hauptman and Karle to advance the science of X-ray crystallography, now a basic tool; quantitative structure-activity relationships have led to the development of at least four commercially successful products (an antibacterial compound, two herbicides, and one fungicide); and insights into molecular structure have been gained from mathematical results in group theory and topology.

In scanning the research needs of theoretical/computational chemistry, the committee found opportunities for synergistic research with almost the entire mathematical sciences community, where that term is used in its broadest sense to include core and applied mathematicians, statisticians, operations researchers, and theoretical computer scientists. Many of the mathematical lines of research that, if reoriented, could contribute to chemistry are already being pursued in other contexts. The matrix in Figure ES.1 displays a subjective assessment of the depth of potential cross-fertilization between major challenges from theoretical and computational chemistry and relevant topics in the mathematical sciences. This matrix is based to some extent on intuition because it is an assessment of future research opportunities, not past results. An "H" in the matrix implies an overlap that appears clearly promising, while an "M" suggests that some synergy between the areas is likely. The absence of an H or an M should not be taken to imply that some clever person will not find an application of that technique to that problem at some point.

Conclusions and Recommendations

As a result of its investigations and collective evaluation of the available information, the committee has reached the following conclusions.

- Several notable "success stories" can be identified, illustrating the value of interdisciplinary stimulation and synergistic research collaboration involving cooperation between the mathematical sciences and the theoretical/computational chemistry communities.
- Many opportunities appear to exist for further collaborations between the mathematical and chemical sciences that could result in high-quality scholarship and research progress that would advance national interests. The productivity of applied computational chemistry would likely be enhanced as a result, which could be potentially significant for industry.
- Active encouragement of further collaborations is warranted because it would likely result in an acceleration of such research progress.
- Cultural differences between the mathematics and the chemistry communities, involving language, training, aesthetics, and research style, have tended to act as barriers to collaboration, even in circumstances that might otherwise suggest the benefit of cooperation.
- Institutional structures and reward systems in the academic community have often placed significant difficulties in the way of collaborative research across traditional disciplinary boundaries, which can be especially inhibiting to those in early career stages.

	Quantum electronic structure	Molecular mechanics	Condensed-phase simulations	Density functionals	N-represent-ability	Design of molecules	Construc-tion of potential energy functions	Gas-phase dynamics	Poly-mers	Topogra-phy of potential energy surfaces	Biological macro-molecules (including protein folding)
Adaptive and multiscale methods	H	M	M				M	M	M	M	M
Special bases	H		M	M				H			
Differential geometry		M				M	M		M	H	H
Functional analysis	H		M	H	H		H	M		M	M
Graph theory	M	M	M			H			M	M	M
Group theory	H		M		M	M		M	M		M
Optimization	H	H	H	H	M	H	H	M	H	H	H
Numerical linear algebra	H	H		M			M			M	
Number theory			M					M		M	
Pattern recognition	M	M	H			H		M	H	H	H
Probability and statistics	H		H		M	H		M	H	H	H
Several complex variables	H		H	M				M		M	M
Topology	M	M	H			H	M	M	H	H	H
Dynamical systems		M	H					H	M	H	M

FIGURE ES.1 Subjective assessment of depth of potential cross-fertilization between major areas of the mathematical sciences and theoretical and computational chemistry. An "H" implies an overlap that is clearly promising and an "M" suggests that some synergy is likely.

- Government funding agencies have for the most part made constructive efforts to identify and fund worthy interdisciplinary and collaborative research. However, this process is still somewhat haphazard. Agencies tend to be organized along traditional disciplinary lines, and the evaluation of interdisciplinary proposals relies on personal contacts between program managers and on timely and comprehensive responses from what is typically a small pool of qualified reviewers. The time lapse involved in the proposal evaluation process thus has often been anomalously long.
- To a large extent, both mathematical scientists and theoretical/computational chemists are relatively unaware of the most exciting recent advances in each others' fields. Consequently both groups tend to be insensitive to the opportunities for interdisciplinary cross-fertilization that could produce intellectual novelty and productivity enhancements on both sides.
- The system of prizes and awards administered by the mathematical sciences and chemistry professional societies is currently not geared to recognize and reward interdisciplinary collaborative research advances.
- The national environment—including Congress, funding agencies, and the professional societies (see, e.g., Joint Policy Board for Mathematics, 1994)—has become perceptibly more conducive to encouraging and supporting interdisciplinary and collaborative research, particularly as it may concern industrial innovation and productivity. Government agencies in particular are currently in a mood to actively encourage joint industrial-academic research, even though proprietary rights barriers to free collaboration are recognized to exist.
- The overwhelming volume of specialized technical literature aggravates the communication problems between fields and occasionally leads to wasted effort, redundancy, and rediscovery. It appears that well-researched and well-written review articles spanning normally disconnected specialties in the mathematical sciences and in theoretical/computational chemistry represent a disproportionately small fraction of the technical literature, in spite of the fact that they can eliminate redundant effort.

In response to these conclusions and to the insights gained from its study, the committee makes the following recommendations:

Undergraduate Education. The best way to attract scientists to interdisciplinary work is to get them interested as undergraduates. It is recommended that universities encourage undergraduate interdisciplinary research courses, seminars, and summer programs.

Graduate Education. Departments in the mathematical and chemical sciences should encourage graduate degrees (both M.S. and Ph.D.) that involve dual (mathematics and chemistry) mentoring. Dual mentoring activity between chemistry and physics and chemistry and biology has been successful in many universities. The committee recommends that mathematics graduate students consider a minor in chemistry instead of a minor in an area of mathematics related to their research specialty. Theoretical and computational chemistry graduate students should consider a minor in mathematics or, alternatively, take a core of mathematical courses appropriate to their interest (perhaps in the framework of a special "interdisciplinary track").

Faculty Interaction. Mathematics and chemistry departments should on occasion invite a person from the other area to speak in a research seminar or a colloquium. Lists of speakers of potential interest to industry should be circulated to local industrial laboratories, and vice versa.

Interdisciplinary Research. The committee recommends that mathematics and chemistry departments encourage and value individual and collaborative research that is at the interface of the two disciplines. Such work has the potential for significant intellectual impact on computational chemistry, and hence on the future evolution of chemical research and its applications to problems of importance in our society.

Professional Societies. Professional meetings in mathematics and chemistry—for instance, those of the American Mathematical Society (AMS), American Chemical Society, Society for Industrial and Applied Mathematics (SIAM), and the Chemical Physics Division of the American Physical Society—would benefit from talks very much like the seminar and colloquium talks described in the recommendation for faculty interaction above, from shorter presentations in special sessions, and from panel discussions. There are already some promising moves in this direction as reflected, for example, by recent AMS sessions on mathematics and molecular biology or SIAM sessions on molecular chemistry problems and optimization. These sessions at national and regional professional society meetings could ultimately lead to focused interdisciplinary meetings.

Prizes and Awards. The committee recommends that professional societies in the mathematical and chemical sciences examine the feasibility of establishing awards and named lectureships for work at the mathematics-chemistry interface. High-level public recognition by peers would be a major step toward breaking down interdisciplinary barriers.

Expository Articles and Books. Professional journals in mathematics and chemistry could enhance their quality, appeal, and influence by publishing expository articles on work at the mathematics-chemistry interface. There is a shortage of books written for someone who is mathematically (chemically) sophisticated and desires fairly precise but nonrigorous chemical (mathematical) explanations.

Interdisciplinary and Industrial Postdoctorals and Sabbaticals. Mathematics and chemistry departments should encourage postdoctoral and faculty sabbatical study at the mathematics-chemistry interface. The committee recommends that the chemical software, pharmaceutical, and chemical industries expand their use of mathematics postdoctorals and faculty on sabbatical leave, and increase their cooperation with and utilization of existing National Science Foundation (NSF) programs such as the University-Industry Cooperative Research Program in the Mathematical Sciences; Industry-Based Graduate Research Assistantships and Cooperative Fellowships in the Mathematical Sciences; Mathematical Sciences University-Industry Postdoctoral Research Fellowships; and Mathematical Sciences University-Industry Senior Research Fellowships. Another opportunity in this regard exists at the Institute for Mathematics and Its Applications at the University of Minnesota, which has an active industrial postdoctoral research program with the aim of broadening the perspectives of recent doctoral recipients in the mathematical sciences and preparing them for research careers involving industrial interaction.

Reference

Joint Policy Board for Mathematics, 1994, *Recognition and Rewards in the Mathematical Sciences,* American Mathematical Society, Providence, R.I.

1
INTRODUCTION

Although much of its discovery process is descriptive and qualitative, chemistry is fundamentally a quantitative science. It serves a wide range of human needs, activities, and concerns, a theme forcefully documented in the comprehensive Pimentel report, *Opportunities in Chemistry* (National Research Council, 1985), which presented the status of chemistry as of 1985. The mathematical sciences provide the language for quantitative science, and this language is growing in many directions as computational science in general continues its rapid expansion. A timely opportunity now exists to strengthen and increase the beneficial impacts of chemistry by enhancing the interaction between chemistry and the mathematical sciences.

Computational chemistry is a natural outgrowth of theoretical chemistry, the traditional role of which involves the creation and dissemination of a penetrating conceptual infrastructure for the chemical sciences, particularly at the atomic and molecular levels. The mathematical sciences have been indispensable allies and have provided vital tools for that role. Theoretical chemistry has also sought to devise and to implement quantitative algorithms for organizing massive amounts of data from the laboratory, and for predicting the course and extent of chemical phenomena in situations that are difficult or even impossible to observe directly; thus, today it is difficult to classify many lines of research as either "theoretical" or "computational." This report tends toward the term theoretical/ computational—any distinction between the two areas is rather misleading because the subject demands both quantitative characterization and conceptual understanding.

Even before the advent of computers as a major component in physical science research, the theoretical tradition in chemistry had accumulated a substantial membership: in its 1966 report entitled *Theoretical Chemistry, A Current Review*, the Westheimer committee estimated that in 1964, approximately 200 theoretical chemists with faculty appointments in graduate-degree-granting institutions could be identified in the United States (National Research Council, 1966, p. 3).

The subsequent three decades have witnessed a revolutionary expansion in the breadth and capability of theoretical and computational chemistry, as well as in its population. These changes, of course, have been driven by the rapid evolution of computers and by their widespread availability in the scientific community. The resulting impact has been enormous and has expanded the range of research activity in theoretical/computational chemistry to encompass the entire spectrum from purely analytical theory, through simulational study of mathematically well-defined models, to the adroit development of powerful and general computational algorithms. Indeed, for the purposes of this document, the committee takes the viewpoint that theoretical/computational chemistry constitutes a seamless continuum of research activities that deserves to be assessed as a whole.

If the mailing lists of theoretical chemistry conferences can be taken as evidence, the current number of theoretical/computational chemists working in the United States has grown to approximately 1000 (John C. Tully, Chairman of 1993 International Conference on Theoretical Chemistry, personal communication). To some extent, this expansion in population has occurred in the academic community. But more significantly, it represents a major growth in the industrial and government sectors, and reflects an increasing realization that theoretical and computational chemistry contributes to the national economic and security welfare. The last three decades have exhibited a general rise in expectations and optimism surrounding the ability of theoretical/computational chemistry to resolve pressing problems both of fundamental scientific character and of clear practical application. The historical record of these expectations can be seen in reports, for example, of workshops and studies held during the early days of the "supercomputer era" (National Research

Council, 1974, 1975, 1976; Schatz, 1984; Berne, 1985). Not surprisingly, physics and engineering manifested similar experiences at the same time (National Science Foundation (NSF) Advisory Committee for Physics, 1981; Lax, 1982; NSF Working Group on Computers for Research, 1983; National Research Council, 1984).

The pervasive significance and widespread applicability of theoretical and computational chemistry may not always be immediately obvious to those not frequently concerned with this activity. Nevertheless, it is central to rational drug design, it contributes to the selection and synthesis of new materials, and it guides the design of catalysts. New quantum mechanical techniques underlie the understanding of electronic properties of materials and have advanced the level of precision at which molecules of at least moderate size can be modeled. Furthermore, computational chemistry software is a set of tools used increasingly by chemists of many persuasions. These various abilities and facilities have proved to be very important to American industry, and their advancement would generate even further industrial benefits. Engaging problems and deep challenges for mathematical scientists are posed by the needs of theoretical and computational chemists, and the products of mathematical research in these areas can have far-reaching ramifications.

The marked growth of theoretical/computational chemistry inevitably has involved a substantial national investment of skilled human resources and of expensive computing resources (both hardware and software). Both of these types of commodities are relatively scarce and are subject to competition between alternative scientific and technological disciplines. Table 1.1 shows, for instance, that software for theoretical and computational chemistry claims much of the cpu usage on the Cray Y-MP at the San Diego Supercomputer Center. Data from other NSF supercomputer centers reveals similar patterns. What Table 1.1 does not show is the heavy dependence of these chemistry codes on mathematical software such as LINPACK and EISPACK. The productivity of these computational resources, broadly construed, must be an issue for continual analysis and informed action by policymakers. In particular, the strong mathematical flavor of theoretical/computational chemistry leads to a natural examination of the efficacy of links between the mathematical and the chemical sciences, and to the past, present, and future roles of interdisciplinary research at the interface between these subjects. These issues constitute basic concerns for the present study.

The 14 chemists, biochemists, and mathematical scientists from industry, government, and academia who attended a 1991 workshop at the National Research Council (NRC) decided that the interface of the mathematical sciences and theoretical/computational chemistry was an area that deserved encouragement, and that a fuller study of the issues was warranted. Subsequently, the Board on Mathematical Sciences and the Board on Chemical Sciences and Technology of the NRC jointly proposed a study to identify research opportunities for the mathematical sciences relevant to computational chemistry, with the goal of engaging the talent of more mathematical scientists in the problems of computational chemistry, which should produce advances of benefit to both the mathematical and the chemical sciences. The phrase "computational chemistry" was to be interpreted to include those areas related to molecular structure and its determination, broadly defined; it was felt that there was less need to promote greater participation by mathematical scientists in the areas of computational chemistry on the macroscopic scale—including such topics as reaction/diffusion modeling and most of chemical engineering. On securing approval and funding for this study, a Committee on Mathematical Challenges from Computational Chemistry was selected, with its first meeting held in Washington, D.C. on March 29-30, 1994. Two subsequent meetings took place: June 9-10, 1994, in Washington, D.C., and September 9-11, 1994, in Woods Hole, Massachusetts.

The statement of task given to the Committee on Mathematical Challenges from Computational Chemistry reads as follows: "The committee will investigate and report on opportunities for collaborative and synergistic research in the mathematical sciences that can accelerate progress in theoretical and computational chemistry and their applications, and make recommendations for promoting this research." It was clear from the outset that the study could not presume to be

TABLE 1.1 Top ten applications in terms of percentage of CRAY C90 usage at the San Diego Supercomputer Center for the period December 1, 1993, to August 17, 1994

Time Used (%)	Application	Description
7.1	ESP	Molecular dynamics
6.7	Gaussian	Quantum chemistry
5.4	AMBER	Molecular dynamics
2.6	TREESPH	Galactic dynamics
2.1	GAMESS	Quantum chemistry
2.0	ARGOS	Molecular dynamics
1.5	CGCM	Coupled ocean-atmosphere global climate model
1.5	DMOL	Quantum chemistry
1.3	COULMETL	Materials science
1.2	DIEL	Materials science

SOURCE: Wayne Pfeiffer, San Diego Supercomputer Center, personal communication.

exhaustive. However, it seemed realistic to strive for representative sampling of the two communities involved and to identify instructive examples of past collaborative successes, likely prospects for interdisciplinary synergy, and barriers to joint research that could be removed or at least lowered.

In order to supplement its own breadth of expertise, as well as to reach out to the mathematical sciences community, the committee invited guests to its first two meetings to learn from their perspectives. At its first meeting, the committee engaged in a lengthy discussion with Richard Herman, chair of the Joint Policy Board for Mathematics, learning about the range of attitudes in that community toward interdisciplinary research and about efforts to adjust the community's priorities on many fronts (Joint Policy Board for Mathematics, 1994). At its second meeting, the committee invited an optimization researcher (Margaret Wright, of AT&T Bell Laboratories, incoming president of the Society for Industrial and Applied Mathematics), a statistician (Douglas Simpson of the University of Illinois at Urbana-Champaign), and a researcher in computational fluid dynamics (David Keyes from the National Aeronautics and Space Administration's Langley Research Center). These guests were invaluable, both for their insights about interdisciplinary research opportunities and for their perspectives on how the committee might influence the mathematical sciences community.

In scanning the research needs of theoretical/computational chemistry, the committee found opportunities for synergistic research with almost the entire mathematical sciences community, where that term is used in its broadest sense to include core and applied mathematicians, statisticians, operations researchers, and theoretical computer scientists in academe, government laboratories, and industry. The common denominator shared by mathematical scientists who have contributed or could contribute to progress in chemistry is not a particular background; rather, it is a willingness to truly collaborate.

Readers may wish to note that two other recently issued reports have a strong bearing on matters considered herein. The NRC has completed a parallel study entitled *Mathematical Research in Materials Science*, which examines many of the same kinds of prospects, barriers, and cures discussed below, although some key distinctions become clear (National Research Council, 1994). The present report gives a somewhat heavier emphasis to biological applications of computational chemistry to avoid excessive overlap with that earlier report. The second is *Recognition and Rewards in the Mathematical Sciences* by a committee of the Joint Policy Board for Mathematics (1994), the recommendations of which are consistent with those contained herein.

The committee believes that this report has relevance and potentially valuable suggestions for a wide range of readers. Several important target audiences and the kinds of benefits they might expect to derive are the following:

1. Graduate departments in the mathematical and chemical sciences could glean suggestions for promising research directions for graduate students and young scientists, ideas about how to foster interdisciplinary collaborations, and insight into new types of job opportunities that may appear in the future.
2. Federal and private agencies that fund research in the mathematical and chemical sciences—including federal policymakers involved in the high-performance computing and communications, materials science, and biotechnology initiatives—can find suggested topics that provide links between the fields, high-priority research topics at the interface, and suggestions for fostering collaborations.
3. Selected industrial and government research and development laboratories can learn of ways in which research from the mathematical sciences could be used to improve the productivity of theoretical and computational chemists.
4. Developers of software and hardware for computational chemistry can gain more insight into the role that the mathematical sciences could play.
5. Selected individual researchers can find inspiration and background for promising research directions (especially for graduate students and young researchers), ways in which their existing lines of research may have parallels or applications in another field, and suggestions for initiating collaborations.

Chapter 2 of this report covers some history of computational chemistry for the nonspecialist, while Chapter 3 illustrates the fruits of some past successful cross-fertilization between mathematical scientists and computational/theoretical chemists. In Chapter 4 the committee has assembled a representative, but not exhaustive, survey of research opportunities. Most of these are descriptions of important open problems in computational/theoretical chemistry that could gain much from the efforts of innovative mathematical scientists, written so as to be accessible introductions to the nonspecialist. Chapter 5 is an assessment, necessarily subjective, of cultural differences that must be overcome if collaborative work is to be encouraged between the mathematical and the chemical communities. Finally, the report ends with a brief list of conclusions and recommendations that, if followed, could promote accelerated progress at this interface. Recognizing that bothersome language issues can inhibit prospects for collaborative research at the interface between distinctive disciplines, the committee has attempted throughout to maintain an accessible style, in part by using illustrative boxes, and has included at the end of the report a glossary of technical terms that may be familiar to only a subset of the target audiences listed above.

References

Berne, Bruce J., 1985, organizer, Supercomputers in the Simulation and Modeling of Chemical Systems, National Science Foundation workshop, Arden House, Harriman, N.Y., April 26–28.

Joint Policy Board for Mathematics, 1994, *Recognition and Rewards in the Mathematical Sciences*, American Mathematical Society, Providence, R.I.

Lax, Peter D., 1982, chairman, Large Scale Computing in Science and Engineering, sponsored by the Department of Defense and the National Science Foundation in cooperation with the Department of Energy and the National Aeronautics and Space Administration.

National Research Council, 1966, *Theoretical Chemistry, A Current Review*, National Academy Press, Washington, D.C.

National Research Council, 1974, *A Study of a National Center for Computation in Chemistry,* National Academy Press, Washington, D.C.

National Research Council, 1975, *A Proposed National Resource for Computation in Chemistry: A User Oriented Facility*, National Academy Press, Washington, D.C.

National Research Council, 1976, *Needs and Opportunities for the National Research for Computation in Chemistry (NRCC)*, National Academy Press, Washington, D.C.

National Research Council, 1984, *Computational Modeling and Mathematics Applied to the Physical Sciences*, National Academy Press, Washington, D.C.

National Research Council, 1985, *Opportunities in Chemistry*, National Academy Press, Washington, D.C.

National Research Council, 1994, *Mathematical Research in Materials Science*, National Academy Press, Washington, D.C.

National Science Foundation Advisory Committee for Physics, 1981, Prospectus for Computational Physics, National Science Foundation, Arlington, Va.

National Science Foundation Working Group on Computers for Research (Kent K. Curtis, Chairman), 1983, *A National Computing Environment for Academic Research*, National Science Foundation, Arlington, Va.

Schatz, George C. (organizer), 1984, Future Directions for Supercomputer Use in Chemistry, National Science Foundation workshop, Evanston, Ill., October 15–17.

2
THE EMERGENCE OF COMPUTATIONAL CHEMISTRY

Computational chemistry has its roots in the early attempts by theoretical physicists, beginning in 1928, to solve the Schrödinger equation (see Box 2.1) using hand-cranked calculating machines. These calculations verified that solutions to the Schrödinger equation quantitatively reproduced experimentally observed features of simple systems such as the helium atom and the hydrogen molecule. Approximate solutions for larger systems and exact solutions to simple model problems allowed chemists and physicists to make qualitative explanations of spectra, structure, and reactivity of all types of matter.

During the Second World War, electronic computers were invented, and in the decade after the war these became available for general use by scientists. At the same time, physicists generally became more interested in nuclear structure and lost interest in the details of molecular structure and spectra. Hence, beginning in the mid-1950s, a new discipline was developed, primarily by chemists, in which serious attempts were made to obtain quantitative information about the behavior of molecules via numerical approximations to the solution of the Schrödinger equation, obtained by using a digital computer. The present success of this field has come largely from the enormous increase in speed, and decrease in cost, of computers, with significant improvements also attributable to many developments in algorithms and methodology. During the 1960s, three major developments in algorithms and methodology made quantum chemistry a useful tool: computationally feasible, accurate basis sets were developed; reasonably accurate approximate solutions to the electron correlation problem were demonstrated; and formulas for analytic derivatives of the energy with respect to nuclear position were derived. These developments were incorporated into several software packages that were made readily available to most chemists in the early 1970s, leading to an explosion in the literature of applications of computations to chemical problems. These programs are used to predict and explain the structure and reactivity of molecules and to complement the information obtained from many types of spectral measurements. Refinement of the program packages has, of course, continued, with emphasis on increased accuracy, increased size of molecules that can be studied, and adaptation to new computer hardware. The present methods have evolved from those that were used to study 1- and 2-atom systems in 1928 through those that were used to study 2- to 5-atom systems in 1970, to the present programs that produce useful quantitative results for molecules with up to 10 to 20 atoms. Much of the current research in new methods is aimed at developing methods that are feasible for even larger molecules.

A classic example of the power of the theoretical/computational approach is the work in the 1960s by W. Kolos and L. Wolniewicz. Explicit r_{12} calculations had been introduced for the hydrogen molecule in 1933 by James and Coolidge, and Kolos and Roothaan, working together in Mulliken's lab, improved these calculations in 1960. Subsequently, Kolos teamed up with Wolniewicz to author a sequence of papers of increasing accuracy. Their results diverged from the accepted (experimentally derived) dissociation energy of H_2. When all known corrections were included, Kolos and Wolniewicz's best estimate of the discrepancy (in 1968) was 3.8 cm^{-1} Thus prodded, experimentalists reexamined the issue and in 1970 a new spectrum of better resolution and a new assignment of the vibrational quantum numbers of the upper electronic state were published. Both of these results were within experimental uncertainty of the best theoretical result.

While the emphasis of one aspect of computational chemistry has been on solving the many-body electronic structure problem, another group of chemists has focused on using the resulting potential energy surface for studying nuclear motion. This has led to a collection of programs for doing

BOX 2.1 The Schrödinger Wave Equation

The time evolution of a (nonrelativistic) quantum mechanical system is prescribed by the Schrödinger wave equation. For a particle with mass m and position $\mathbf{r}$ that is moving under the influence of a potential $V(\mathbf{r})$, this equation reads:

$$-(\hbar/i)\ \partial\psi(\mathbf{r},t)\ /\ \partial t = \mathbf{H}\ \psi(\mathbf{r},t)\ ,$$

where $\mathbf{H}$ is the linear Hermitian Hamiltonian operator:

$$\mathbf{H} \equiv -(\hbar^2/2m)\ \nabla^2 + V(\mathbf{r})\ .$$

Here $\hbar$ is Planck's constant divided by 2π and has the very small value 1.05×10^{-27} erg-second characteristic of the submicroscopic quantum regime. The wavefunction ψ generally is complex; its amplitude squared $|\psi|^2$ provides the probability distribution for the position of the particle at time t. The linearity of the Schrödinger wave equation and the Hermiticity of $\mathbf{H}$ guarantees conservation of probability.

Time-independent solutions to the wave equation that have physical significance fall into two categories and require two distinct boundary conditions. The first ("bound states") are square-integrable eigenfunctions of $\mathbf{H}$; they are bounded, vanish at infinity, and provide a discrete spectrum of real energy eigenvalues for $\mathbf{H}$. The second ("scattering states") are a continuum of solutions that lie in energy above the infinite-r limit of $V(\mathbf{r})$ (if it exists), and the wavefunction remains nonzero (and oscillatory as a plane wave or sinusoidal function) in this asymptotic limit.

Quantum mechanical phenomena implied by the Schrödinger wave equation have a strongly counterintuitive flavor. The discreteness of bound-state energies contrasts starkly with the continuous energies available to a dynamical system in classical mechanics. The corresponding discreteness of the energy spectrum of transitions between pairs of quantum bound states has spawned the colloquially overused and misused phrase "quantum leap." In addition, the wavefunction can be nonvanishing in regions where V exceeds the total energy, leading to the well-known, but often mystifying, tunneling phenomenon; this permits particles to pass through potential energy barriers that in classical mechanics would provide absolute blockage. Less well known, but equally mystifying, is the reverse phenomenon originally pointed out by von Neumann and Wigner (1929): some potential functions $V(\mathbf{r})$ have the capacity to trap particles in square-integrable eigenstates at energies above the absolute maximum of V, again in contradiction to classical mechanical behavior.

The Schrödinger wave equation adapts naturally to the description of many mutually interacting particles. The case most prominently considered in chemistry is that of two or more electrons, where V includes Coulombic interactions for electron-electron repulsions and electron-nucleus attractions. However, it is important to account for the fact that electrons in particular are endowed with intrinsic angular momentum, denoted by "spin," that is quantized to display only two allowed values, "up" and "down." In the strictly nonrelativistic regime (chemically, that which involves only atoms with low atomic numbers), these electron spins are invariants of motion and can be formally eliminated from the mathematical problem, provided that the remaining spatial wavefunction for the electrons satisfies appropriate symmetry conditions. In the case of two-electron systems (examples are the helium atom and the hydrogen molecule), the spatial wavefunction must either be symmetric under interchange of electron positions (if one "up" spin and one "down" spin are present) or antisymmetric (if both spins are "up" or both are "down"). Analogous, but more complicated, interchange conditions are applicable to the spatial wavefunction when more electrons are present.

Reference

von Neumann, J., and E. Wigner, 1929, *Phys. Z.* 30:465.

classical, semiclassical, and quantum calculations. Since 1980, use of these programs has become a routine tool for modeling molecules and gas-phase chemical reactions. These computations yield collision cross sections, both differential and integral, for elastic, inelastic, and reactive events. These approaches require, as with transition state theory, potential energy surface(s) obtained using quantum chemical methods of solution of the electronic Schrödinger equation. The Schrödinger equation for nuclear motion is solved subject to a scattering boundary condition, which takes the form of coupled differential, integro-differential, algebraic, or integral equation systems. The methods used to solve these coupled systems of equations are drawn from the applied mathematics literature as well as from algorithm improvements developed by computational chemists.

Meanwhile, simpler approximations have long been used by chemists to estimate the energy of molecules near their equilibrium geometry. In the molecular mechanics approach (see Box 2.2) the total energy of a chemical system is approximated by a sum of simple terms involving distances between atoms, bond angles, and dihedral angles. These terms involve estimated parameters that are assumed to have the same values as similar parameters obtained by data fitting for simpler molecules. (Chemists have long known that many structural and energetic features of molecules are nearly transferable between similar subfragments of molecules.) This representation of the energy has made possible the modeling of biological systems and rational drug design. It is also at the heart of the computational engine of many programs that produce three-dimensional computer graphics images of molecules. Molecular mechanics has become so prevalent that many chemists now equate it with computational chemistry. This approach has allowed the modeling of molecules with thousands of

BOX 2.2 Molecular Mechanics/Molecular Dynamics

Molecular mechanics and molecular dynamics refer to methods for computing certain molecular properties, particularly molecular structure and relative energy. They both typically use fairly simple potential energy functions that are derived from classical mechanics (e.g., a parabolic function to calculate the energy required to stretch or to compress a chemical bond). In addition, they both rely on parameters that are derived either from experiment (e.g., infrared spectroscopy and X-ray crystallography) or from quantum mechanics-based calculations (e.g., high-level ab initio molecular orbital calculations). A collection of potential energy functions and the associated parameters that are employed for molecular mechanics/molecular dynamics calculations is frequently referred to as a "force field"; thus, calculations that utilize the molecular mechanics or molecular dynamics approach are often referred to as empirical force field calculations.

The molecular mechanics method is generally employed to compute the relative energies of different geometries (conformations) of the same molecule that arise from rotations about chemical bonds as well as relative energies of intermolecular complexes. Often, energy minima are sought; thus, the molecular mechanics method is frequently coupled with optimization procedures. On the other hand, in the molecular dynamics method, Newton's equations of motion are solved by using the gradient of the above-mentioned potential energy function (force field) to compute the dynamic trajectory of a molecule or of an ensemble of molecules. Both the molecular mechanics and the molecular dynamics methods have found widespread use in the modeling of biomolecular systems, for which quantum mechanical calculations are simply not practical due to the overwhelming number of particles involved. Nonetheless, these methods are quite accurate for the estimation of certain molecular properties (i.e., those for which classical mechanics is appropriate), and they have been successfully employed to compute conformational energies (as described above), to estimate the binding affinity of small molecules bound to a macromolecular receptor, and as an adjunct for the refinement of structures derived from protein X-ray crystallography and protein nuclear magnetic resonance spectroscopy.

BOX 2.3 Chemist, Mathematician, or Physicist?

Douglas R. Hartree was an example of a truly interdisciplinary scientist. He invented the self-consistent field approach in 1927 to approximate the solution of the Schrödinger equation and did a number of calculations on atomic structure. His undergraduate studies had been interrupted by the First World War, during which he did ballistics computations. After completing his Ph.D., he held three chairs, first in Applied Mathematics at Manchester, then in Theoretical Physics at Manchester, then in Mathematical Physics at Cambridge. In 1934, he built the first differential analyzer in England. From 1939 until 1945, he used this in support of the war effort. After the war he was a consultant on the ENIAC computer and developed some of the first computer-based methods for solving partial differential equations. He taught numerical analysis at Cambridge University and wrote an insightful text on the subject, first published in 1952.

atoms. The practical disadvantage is that only structural types previously encountered in smaller molecules can be parameterized for larger molecules, so many parameters remain unknown. The conceptual disadvantage is that this is no longer a first principles theory and the connection to the Schrödinger equation is unclear. Hence, there can be no rigorous estimate of the potential errors in this approach and its success relies on chemical intuition for finding suitable molecules from which to develop the "transferable" parameters.

Another important thread in theoretical chemistry has been the study of many-particle systems such as liquids, solid materials, and biological macromolecules. The major framework for this study has been statistical mechanics—a subject with its formal roots in the nineteenth century. In the 1930s, the study by physical chemists of structure and thermodynamics accelerated with the advent of simple ideas about intramolecular and intermolecular forces. Equilibrium statistical mechanics has offered many questions of principle—for example, the question of the nature, and even definition, of phase transitions. These questions fostered a long-standing cross-fertilization between workers in both the mathematical and chemical communities (see Box 2.3). Similarly the study of phenomena away from equilibrium (e.g., the transport phenomena of hydrodynamics and the chemical rates) attracted the fundamental thinkers in statistical mechanics starting in the 1950s. Recently, corresponding deep questions of principle about disordered systems such as glasses have attracted workers from both communities.

Although a large part of statistical mechanics can be studied without computers, machine calculations for many-body simulation made an early impact in the 1950s and have grown to be the dominant mode of investigation. Monte Carlo methods, invented at the weapons laboratories by workers such as Fermi, Ulam, von Neumann, Metropolis, and Teller, were used immediately to address the many-body problems relevant to the thermodynamics of liquids. Such Monte Carlo approaches were adapted quickly to the study of polymers as well. The numerical solution of Newton's laws for many-particle models, so-called molecular dynamics (see Box 2.3), was also first carried out by theoretical chemists in the late 1950s and early 1960s. The application of molecular dynamics and Monte Carlo methods to proteins and other biomolecules in the 1970s has led to their widespread use throughout the theoretical and experimental chemical communities. Since significant advances in the efficiency of the algorithms used in molecular dynamics and Monte Carlo simulation are needed to address the forefront questions such as protein folding, a renewed contact of theoretical chemists with the numerical mathematics community has recently involved collaborative efforts of mathematicians, chemists, and physicists.

The advent of molecular quantum mechanics was followed by a very successful theory of chemical reaction rates that modeled a reactive event as passage over a reaction barrier on a multidimensional potential energy surface representing the energy as a function of the internal coordinates of the reacting system. In its simplest form, the model corresponds to the system moving from reactants to transition states (the critical configuration), from which the system moves to reaction products. This conceptually simple model has remained the predominant approach for estimating rates of chemical reactions. Because of the multidimensionality of the reactive system, however, it is computationally difficult to implement rigorously. Over the years, efforts have focused on improving methods to estimate reaction barriers and properties of the reactants, and these have required better solutions of the electronic and nuclear transition states.

The roots of much of the mathematics now finding application to computational chemistry extend back at least to the eighteenth or nineteenth century, although, as illustrated in Chapters 3 and 4 of this report, the most up-to-date developments in the mathematical sciences can also be very natural tools. Group theory traces its origin to fundamental studies of geometries, but from it has come the theory of groups of motions, continuous groups, Lie groups, and Lie theory. The need to understand functions on the sphere and other surfaces led to the representation theory of groups and to various kinds of function theory. These theories grew up with the creation of quantum mechanics and fed, and were fed by, quantum mechanics. Much of operator theory and integral equations came from physics and engineering, as did the general theory of harmonic analysis. Numerical linear algebra and numerical analysis developed largely as tools for fluid mechanics and military applications, but their usefulness is vastly more widespread than that.

After World War II the mathematics community entered a period of intense development of its core, accelerating the growth of fields such as topology, number theory, algebraic geometry, and graph theory. Advances were largely motivated by questions generated by the internal structure of mathematics and not by contact with the outside world. In recent years, however, attention has once again turned outward, and the products of this intense period are now being applied widely in novel ways. The advent of modern computing capacity has enabled mathematicians to generate computational algorithms that yield answers—when combined with proper modeling techniques—to important practical problems. Success has been achieved in signal processing, sound and image compression, flow problems, and electromagnetic theory. Historically, mathematical scientists have worked more closely with engineers and physicists than with chemists, but recently many fields of mathematics such as numerical linear algebra, geometric topology, distance geometry, and symbolic computation have begun to play roles in chemical studies.

Before proceeding to accounts of past and potential contributions that mathematics can make to progress in chemistry, it should be emphasized that the challenge of interdisciplinary research is not one of scientific content alone, but also one of scientific process. Neither the chemist nor the mathematician is generally the optimal person to construct a mathematical model, as the model by its very nature lies at the interface between theory and observation. To build the model, an iterative process of refinement is required, in which mathematical considerations motivate approximations that need to be checked against reality, and in which key chemical insights necessarily force levels of mathematical complexity. It is exactly this need for iterative model construction that may motivate the collaboration of mathematicians and chemists, against the self-referential and conservative tendencies of each discipline. Focusing on this process of iterative model construction can help clarify the roles of the collaborators in interdisciplinary research, and by extension illustrate the goals for their respective disciplines as attempts are made to lower the hurdles to such collaborations. The model is both the interface between the disciplinary boundaries and the lingua franca between the cultures.

3
EXAMPLES OF CONSTRUCTIVE CROSS-FERTILIZATION BETWEEN THE MATHEMATICAL SCIENCES AND CHEMISTRY

Use of Statistics to Predict the Biological Potency of Molecules Later Marketed as New Drugs and Agricultural Chemicals

Because the search for new drugs or pesticides typically involves the investigation of thousands of compounds, many research investigators have sought computer methods that would correctly forecast the biological properties of compounds before their synthesis. Box 3.1 describes how searches for new drugs or pesticides are done. There are four well-documented cases of the use of computer methods, particularly quantitative structure-activity relationship (QSAR) methods, as an integral part of the design of compounds that are now marketed as drugs or agrochemicals. Not only are these compounds commercial successes for the companies that developed them, but they benefit mankind by aiding in the treatment of disease or increasing the food supply. This viewpoint has been so successful that recently a company, Arris, was founded to incorporate the direct involvement of mathematicians in the development of proprietary drug design software.

The Hansch-Fujita QSAR method (Hansch and Fujita, 1964) was developed in the early 1960s and has become widely used by medicinal and agricultural chemists. In this method, one first describes each molecule in terms of its physical properties and then uses statistical methods to uncover the relationship between some combination of these physical properties and biological potency.

Usually in QSAR methods the relationships are examined with multiple linear or nonlinear regression, classical multivariate statistical techniques. However, discriminant analysis, principal components regression, factor analysis, and neural networks have been applied to these problems as well. More recently, the partial least squares (PLS) method (Wold et al., 1983) has found wide use in both QSAR and analytical chemistry. Although PLS was originally developed by a statistician for use in econometrics, its widespread utility in chemistry has prompted additional statistical research to improve its speed and its ability to forecast the properties of new compounds, and to provide mechanisms to include nonlinear relationships in the equations.

Recently, Boyd described four cases in which QSAR and other computer analysis led to a commercial product (Boyd, 1990). He documented each case carefully by correspondence with the original inventors. The first is the antibacterial compound norfloxacin marketed for human therapy in Japan, the United States, and other countries. It is up to 500 times more potent than previously marketed compounds of this class. Additionally, it is effective against *Pseudomonas*, a difficult organism to control. Norfloxacin and its subsequent derivatives achieve a clinical efficacy of approximately 90%. Norfloxacin was designed at the Kyorin Pharmaceutical Company in Japan from a traditional QSAR analysis that used regression analysis of about 70 compounds.

The second and third QSAR-designed molecules to reach the market are both herbicides. Metamitron, discovered by Bayer AG in Germany, was based on a QSAR that involved the multiple linear regression analysis of 22 compounds. In 1990 it was the best seller in Europe for the protection of sugar beet crops. The other herbicide, bromobutide, has been marketed in Japan since 1987. It was developed at Sumitomo Chemical Company in Japan based on QSAR analysis of 74 compounds.

The final example concerns the fungicide myclobutanil, which entered the European market in 1986 for the treatment of grape diseases and was introduced to the U.S. market in 1989. It was developed by Rohm and Haas in the United States. The design of myclobutanil involved traditional

BOX 3.1 Rational Drug Design

The traditional discovery of new drugs is an empirical process that starts with a compound of marginal biological activity. This "lead" compound either is discovered serendipitously by the random screening of a large number of compounds (often obtained from libraries of previously synthesized molecules) or is obtained by preparing analogues of a natural ligand (i.e., a small molecule such as a hormone that binds to a biomacromolecule such as an enzyme). Chemical intuition and experience as well as ease of synthesis serve to suggest other closely related molecules (analogues) for synthesis. This process is iterative and continues until a compound is discovered that not only possesses the requisite activity toward the target but also possesses minimal activity toward other biomacromolecules (i.e., it is selective and nontoxic). The compound must also have a desirable duration of action in a suitable dosage form, its synthesis must not be too costly so that its use will be cost-effective, it must be patentable, etc. This process can take many years, can cost millions of dollars, and often does not result in a marketed product. Any method that would make this process more efficient is clearly useful. Thus, chemists in the pharmaceutical industry have sought a more rational procedure for the discovery and design of new drugs.

If the three-dimensional structure of the target biomacromolecule has been determined (e.g., by using X-ray diffraction or nuclear magnetic resonance spectroscopic techniques), a technique that has been termed structure-based drug design can be used for the design of new molecules with the potential to become useful therapeutic agents. If the three-dimensional structure of the target is unavailable, then a hypothetical model is formulated with the goal of describing the molecular features required if a particular compound is to elicit a desired biological response. This model, of course, can be validated only after a number of compounds have been synthesized and tested for their biological activity so that a statistical relationship between biological activity and physical molecular properties (i.e., a quantitative structure-activity relationship, or QSAR) can be established. Nonetheless, such a model is highly useful for focusing the synthetic effort on compounds that have the greatest chance of exhibiting increased biological activity. Rational drug design is heavily dependent on computational chemistry techniques, and advances in rational drug design are tightly coupled to advances in new algorithms for computer-assisted molecular modeling.

To design a new ligand for a biomacromolecule of interest using the three-dimensional structure of the target biomacromolecule as a guide, the structure of the target must have been found with sufficient resolution to be of utility. One must then attempt to predict the bound geometry and intermolecular interactions responsible for the high binding affinity of novel potential ligands (or molecular fragments) associated with the biomacromolecular target. Computer algorithms have been developed over the past few years that aid in the identification of potential docking modes. These algorithms have also been used to identify, from three-dimensional databases, molecules that can potentially dock (and hence bind) to a biomacromolecular target. The prediction of biological activity of a potential ligand prior to synthesis represents another essential activity for the structure-based design of new drugs. This endeavor represents an enormous challenge for structure-based drug design, but some progress has been made using statistical mechanics-based free energy perturbation techniques that involve computer simulations employing molecular dynamics or Monte Carlo methods or by using QSAR methods that rely on the three-dimensional properties of the bound ligand.

In spite of the obstacles associated with employing an analytical approach to the design of new drugs, rational drug design has, nonetheless, been of enormous utility to the pharmaceutical industry. The QSAR method has played a role in the development of a number of drugs currently undergoing clinical trials and there are marketed products for which QSAR has been instrumental. A number of potential drugs that have been discovered using structure-based drug design techniques are currently under preclinical or clinical investigation for the treatment of diseases that include cancer, AIDS, rheumatoid arthritis, psoriasis, and glaucoma.

QSAR on 67 compounds and three-dimensional molecular modeling to explain the QSAR and to provide a model of how the compounds bind to their biological target.

Although these successes are real accomplishments, researchers in medicinal and agricultural chemistry would like to extend the methods to more cases; such extensions create an opportunity for creative mathematical insights. The thrust of new research in QSAR has been to calculate the descriptors of the molecules from their three-dimensional arrangements of atoms and electrons in space (Kubinyi, 1993). The problem is that one of the popular methods, the comparative molecular field analysis (CoMFA), generates thousands of descriptors for each molecule, whereas the dataset typically contains biological activity for only 10 to 30 members. While partial least squares can properly handle such data, it is sensitive to random noise, with the result that the true signal may be masked by irrelevant predictors. QSAR workers need a new method to analyze matrices with thousands of correlated predictors, some of which are irrelevant to the end point. This is an opportunity for a mathematical scientist to contribute an original approach to an important problem.

The new company Arris was founded on the basis of a close collaboration of mathematicians and theoretical chemists. They have produced QSAR software that examines the three-dimensional properties of molecules using techniques from artificial intelligence (Jain et al., 1994). The initial results from this work are promising and suggest that further improvements in three-dimensional QSAR could result from additional collaborations between mathematicians and theoretical chemists.

References

Boyd, D.B., 1990, *Successes of Computer-Assisted Molecular Design*, Reviews in Computational Chemistry, VCH Publishers, New York, pp. 355–371.

Hansch, C., and T. Fujita, 1964, Rho sigma pi analysis: A method for the correlation of biological activity and chemical structure, *Journal of the American Chemical Society* 86:161–162.

Jain, A., K. Koile, and D. Chapman, 1994, Compass: Predicting biological activities from molecular surface properties; Performance comparisons on a steroid benchmark, *Journal of Medicinal Chemistry* 34:2315–2327.

Kubinyi, H., ed., 1993, *3D QSAR in Drug Design: Theory, Methods, and Applications,* ESCOM, Leiden.

Wold, S., H. Marten, and H. Wold, 1983, The multivariate calibration problem in chemistry solved by the PLS method, *Matrix Pencils (Lecture Notes in Mathematics)*, Springer-Verlag, Heidelberg, pp. 286–293.

Numerical Analysis

Since much of current computational chemistry is based on numerical computation, it is not surprising to find successful transfers of information from the numerical analysis community to computational chemistry. Subroutine packages such as LINPACK, EISPACK, and those in the NAG and IMSL libraries codify algorithms for solving linear equations and eigenproblems, developed by the numerical linear algebra community over a period of decades. These software packages provided reliable and well-documented solutions to common mathematical problems with a fixed and well-defined user interface. The internal details of these components have been enhanced from time to time, for example, through the use of the basic linear algebra subroutines (BLAS) that allowed computer vendors to optimize performance to some extent without requiring significant changes to the source code of these packages. This allowed them to be used effectively on vector supercomputers when they were introduced. Many of the EISPACK routines were incorporated into widely used quantum chemistry packages such as GAMESS, HONDO, MELD, and COLUMBUS when these packages were "vectorized."

Recent advances in parallel computing have allowed much larger problems to be addressed cost-effectively. The LAPACK project funded by the National Science Foundation (NSF) has developed software for solving linear equations on modern high-performance computers (Demmel et al., 1993). The Defense Department's Advanced Research Projects Agency is funding a similar project called PRISM to develop scalable implementations of an eigensolver based on the invariant subspace decomposition approach (ISDA), as well as parallel implementations of fundamental linear algebra operations such as band reduction, tridiagonalization, and matrix multiplication.

A more recent advance in numerical analysis is the method of multipole expansions for computing long-range forces, such as Coulombic forces, more efficiently through the use of very accurate simplified approximations (Draghicescu, 1994) in the far field. This has been applied successfully in molecular dynamics with implementations for both sequential and parallel computers (Ding et al., 1992a,b).

Many computational chemistry codes have been adapted to work efficiently on parallel computers by a variety of techniques. Much of this has been achieved by modifying programs and data structures using techniques from computer science (Plimpton and Hendrickson, 1994). However in other codes, new parallel algorithms have been developed. For example, grid-based electrostatics

BOX 3.2 Research Opportunities in Parallel Computing

Considerable opportunities for advances remain regarding parallel computation that could impact computational chemistry. For example, multigrid methods solve grid-based electrostatics problems in an optimal amount of work (and storage) on sequential computers. That is, the unknown potential can be determined on a grid of n points in $O(n)$ work and storage. However, the standard adaptation of multigrid methods to parallel computers is not optimally efficient. A significant theoretical problem is whether a solution technique exists that uses only $O(n/p)$ work on a parallel computer with p processors.

calculations (Davis et al., 1990) provide one of the most difficult simulations to parallelize with full efficiency, due to the long-range interactions implicit in such partial differential equations (see Box 3.2). No part of the problem domain can be treated independently of any other, and so there is no natural parallelism in such problems. Like many algorithms in scientific computation, parallelism must be created at the expense of communication between processors that would be absent in the uniprocessor implementation.

Despite the inherent difficulties in parallelizing such problems, novel domain decomposition methods have provided effective parallel iterative methods (Ilin et al., 1995). These techniques allow extremely large problems to be solved in a moderate amount of time on massively parallel processors. They have been incorporated into the computer code UHBD, which has been used effectively to study biomedically significant enzymes in different ways, providing critical insight into the discovery of new behavior (Gilson et al., 1994) and even allowing the engineering of new, more effective enzymes (Getzoff et al., 1992).

References

Davis, M.E., J.D. Luty, B.A. Allision, and J.A. McCammon, 1990, Electrostatics and diffusion of molecules in solution: Simulations with the University of Houston Brownian Dynamics program, *Computer Physics Communications* 62:187–197.

Demmel, James W., Michael T. Heath, and Henk A. van der Vorst, 1993, Parallel numerical linear algebra, *Acta Numerica* 3:111–197.

Ding, Hong-Qiang, Naoki Karasawa, and William A. Goddard III, 1992a, The reduced cell multipole method for Coulomb interactions in periodic systems with million-atom unit cells, *Chemical Physics Letters* 196:6–10.

Ding, Hong-Qiang, Naoki Karasawa, and William A. Goddard III, 1992b, Atomic level simulations on a million particles: The cell multipole method for Coulomb and London nonbond interactions, *J. Chemical Physics* 97:4309–4315.

Draghicescu, C.I., 1994, An efficient implementation of particle methods for the incompressible Euler equations, *SIAM J. Numer. Anal.* 31:1090–1108.

Getzoff, Elizabeth D., Diane E. Cabelli, Cindy L. Fisher, Hans E. Parge, Maria Silvia Viezzoli, Lucia Banci, and Robert A. Hallewell, 1992, Faster superoxide dismutase mutants designed by enhancing electrostatic guidance, *Nature* 358:347–351.

Gilson, M.K., T.P. Stratsma, J.A. McCammon, D.R. Ripoll, C.H. Faerman, P.H. Axelsen, I. Silman, and J.L. Sussman, 1994, Open "back door" in a molecular dynamics simulation of acetylcholinesterase, *Science* 263:1276–1278.

Ilin, A., B. Bagheri, L.R. Scott, J.M. Briggs, and J.A. McCammon, 1995, Parallelization of Poisson-Boltzmann and Brownian Dynamics calculation, in *Parallel Computing in Computational Chemistry*, T.G. Mattson, ed., ACS Books, Washington, D.C.

Plimpton, S., and B. Hendrickson, 1994, *A New Parallel Method for Molecular Dynamics Simulation of Macromolecular Systems*, Report 94-1862, Sandia National Laboratories, Albuquerque, N. Mex.

Distance Geometry

The idea of modeling complex molecules by using residue-residue cartesian distances as a guide for understanding the nature of protein folding and energetics stimulated work in the early 1970s (Kuntz, 1975; Liljas and Rossman, 1975). It was clear, though, that more mathematical machinery was needed. The area of distance geometry already existed in the work of Blumenthal (1970), while closely related mathematics, called multidimensional scaling, was developed by Kruskal and Wish (1978) and incorporated into advanced statistical packages. In essence, distance geometry is a method to work in spaces with greater than three dimensions, allowing distance constraints to be satisfied that could not be satisfied in three dimensions. Distance geometry helps one move from collections of distances between points to possible coordinates for these points. It also helps one distinguish important information from the standard restrictions imposed on us by living in three dimensions.

Tools were developed to go from upper and lower bound distances to three-dimensional structures, a process that required projecting an object from many-dimensional space into three. A seminal paper explored what distance information was needed to determine a three-dimensional structure to a given resolution (Havel et al., 1979), and later work concluded that a large amount of imprecise data could be sufficient to determine a macromolecular structure to a high resolution (Havel et al., 1983). This was the time when nuclear magnetic resonance (NMR) spectroscopists were beginning to be able to extract atom-atom distances for matter in solution, which could then be compared with those that could be found in the crystal by X-ray diffraction (Havel and Wuthrich, 1985). Distance geometry or

BOX 3.3 Protein Microtutorial

Proteins are large molecules composed of smaller chemical units (residues), the 20 naturally occurring amino acids, stably linked together in an ordered linear sequence. The amino acids possess distinct sizes, shapes, and mutual interactions. Consequently, the ordered sequence (denoted by "primary structure" of the protein) controls physical and biological properties. In its simplest form, "the protein folding problem" consists in predicting the three-dimensional stable shape of protein molecules from the primary structure, to facilitate understanding those physical and biological properties.

While the fixed chemical structure of proteins constrains most of their chemical bond lengths and the angles between neighboring chemical bonds to lie within tight limits, a substantial number of degrees of freedom describing the local amino acid packing ("secondary structure") and the overall three-dimensional shape of the protein (its "tertiary structure" or "conformation") remain to be determined. It is believed that the conformation holds the key to physical and biological properties. An experimentally important type of information relevant to a protein's preferred conformation is a set of distances between selected amino acids.

The folded conformation of a protein is analogous to the posture of a human body. Body motion at skeletal joints plays the role of protein backbone and amino acid side chain folding degrees of freedom. Given a selected set of distances between skeletal joints, for instance, one might be able to infer what posture the human subject had adopted or at least to rule out several possibilities. Similarly, distance geometry for proteins can be used to reconstruct their three-dimensional conformation.

the related approach to refine the molecule in real space (Braun and Go, 1992) turned out to be useful methods to turn NOE (nuclear Overhauser effect) distances into three-dimensional structures. Distance geometry has continued to be a key tool in the NMR spectroscopist's arsenal, providing not only the structures, but also a quantitation of how accurately they are known.

Distance geometry is an important technique in computational chemistry. The focus of the original work was to predict protein structure from amino acid sequence (see Box 3.3) and work continues along these lines using residue-residue potentials (Maiorov and Crippen, 1992). The use of distance geometry in NMR structure determination is mentioned above. Distance geometry has also been used as a tool in the development of QSARs in macromolecule-ligand binding (Ghose and Crippen, 1990) and in a docking procedure to find different orientations that ligands can have when bound to macromolecules (Kuntz et al., 1982).

An example of this last use of distance geometry is its application to the intermolecular docking of a small molecule to a protein, similar to what was done to produce the cover illustration. Distance geometry is applied by setting the distances between interacting atoms to their ideal intermolecular distance (in contrast to the bond length). The result is a general program for solving conformational problems involving one or more molecules with implicit interatomic constraints taken from the given molecular structure and explicit distance constraints added by the user. A program by Blaney et al. (1990), for example, allows one to solve model building problems on complex molecular systems that are very difficult or impossible to solve by conventional methods.

Distance geometry has also been used to establish which, if any, three-dimensional conformations of a set of molecules can be superimposed. One treats the whole ensemble of molecules simultaneously with inter-molecular distances set to zero if the atoms are

Docking is the fitting and binding of small molecules (ligands) at the active sites of biomacromolecules (e.g., enzymes and DNA).

to be superimposed or to infinity if they need not be. This has become a useful method for analyzing three-dimensional structure-activity relationships (Sheridan et al., 1986).

In summary, distance geometry is a general and powerful tool for creating three-dimensional structures, usually by going into a higher-dimensional space and then projecting into three dimensions. Its power lies in exploring "conformational" space (the universe of all possible spatial arrangements of a molecule) and assessing how convincingly the data (often experimental) have implied the structure. It has been applied in this guise to structures from small organic ring systems to proteins in the ribosomal machinery. It is clearly an area in which a fundamental technique from mathematics was brought to bear in important areas of structural chemistry and biochemistry (Crippen and Havel, 1988).

Given the importance of all areas in which distance geometry has been applied to date (protein folding, ligand docking, conformational analysis), future development in this area is likely to be important for computational chemistry (Crippen, 1991). Some of the remaining major mathematical research challenges in distance geometry applied to chemistry include (1) the need to develop improved sampling algorithms (e.g., partial metrization); (2) the need for practical algorithms to solve tetrangle and higher-order inequalities; (3) the need to develop biased sampling approaches that avoid previously sampled configurations; and (4) energy embedding—given a pairwise potential, how can one best solve for the global minimum in $N-1$ space and then "squeeze" the system down to 3-space? A perspective on some of these challenges is given in recent reviews (Crippen, 1991; Blaney and Dixon, 1994).

References

Blaney, J.M., and J.S. Dixon, 1994, Distance geometry in molecular modeling, in *Reviews in Computational Chemistry*, vol. 5, K.B. Lipkowitz and D.B. Boyd, eds., VCH Publishers, New York, pp. 299-335.

Blaney, J.M., G.M. Crippen, A. Dearing, and J.S. Dixon, 1990, DGEOM program #590, Quantum Chemistry Program Exchange, Bloomington, Ind.

Blumenthal, C.M., 1970, *Theory and Applications of Distance Geometry*, Chelsea Publishing Co., Bronx, N.Y.

Braun, W., and N. Go, 1985, Calculation of protein conformations by proton-proton distance constraints: A new efficient algorithm, *J. Mol. Biol.* 186:611–626.

Crippen, G.M., 1991, Chemical distance geometry—current realization and future projection, *J. Math. Chem.* 6:307-324.

Crippen, G.M., and T.F. Havel, 1988, *Distance Geometry and Molecular Conformation*, Research Studies Press, John Wiley & Sons, New York.

Ghose, A.K., and G.M. Crippen, 1990, Modeling the benzodiazepine receptor binding site by the general three-dimensional quantitative structure-activity relationship method REMOTEDISC, *Mol. Pharm.* 37:725–734.

Havel, T.F., and K. Wuthrich, 1985, An evaluation of the combined use of nuclear magnetic resonance and distance geometry for the determination of protein conformations in solution, *J. Mol. Biol.* 182:281–294.

Havel, T.F., G.M. Crippen, and I.D. Kuntz, 1979, Effects of distance constraints on macromolecular conformation. II. Simulation of experimental results and theoretical predictions, *Biopolymers* 18:73–81.

Havel, T.F., I.D. Kuntz, and G.M. Crippen, 1983, The theory and practice of distance geometry, *Bull. Math. Biol.* 45:665–720.

Kruskal, J.B., and M. Wish, 1978, *Multidimensional Scaling,* Sage Publishing, Beverly Hills, Calif.

Kuntz, I.D., 1975, An approach to the tertiary structure of globular proteins, *J. Am. Chem. Soc.* 97:4362–4366.

Kuntz, I.D., J.M. Blaney, S.J. Oatley, R. Langridge, and T.E. Ferrin, 1982, A geometric approach to macromolecule-ligand interactions, *J. Mol. Biol.* 161:269–288.

Liljas, A., and M. Rossman, 1975, Recognition of structural domains in globular proteins, *J. Mol. Biol.* 85:177–181.

Maiorov,V.M., and G.M. Crippen, 1992, Contact potential that recognizes the correct folding of globular proteins, *J. Mol. Biol.* 227:876–888.

Sheridan, R.P., R. Nilakantan, J.S. Dixon, and R. Venkataraghaven, 1986, The ensemble approach to distance geometry: Application to the nictotinic pharmacophore, *J. Med. Chem.* 29:899–906.

Mathematics and Fullerenes

The structures and properties of the fullerene molecules—"buckyballs" (see Figure 3.1) and related highly symmetric carbon molecules that are roughly spherical—have been linked with some very central areas of mathematics.[1] Topology can provide insights into the types of such structures that can and cannot exist; the symmetries of the molecules, which underlie some of their interesting properties, are naturally described with group theory; and graph theory can give insight concerning the vibrational modes of such molecules.

The prototypical buckyball molecule, C_{60}, is composed of 60 carbon atoms linked into a shape reminiscent of a soccer ball, mathematically known as a truncated icosahedron. Other fullerenes that have been observed are composed of more than 60 carbons, except for one member of the family that contains only 44. The "surfaces" of all members are composed solely of pentagons and hexagons and share the property that each vertex connects exactly three edges. (This latter property follows from the chemical bonding of carbon atoms.) Such polyhedral surfaces are subject to a classic topological relationship derived by Euler:

$$\sum_n (6-n) f_n = 12 ,$$

where the summation is over all faces of the polyhedron and f_n is the number of faces with n sides. This expression leads immediately to a property observed in all the fullerenes observed so far in the laboratory: since n is found experimentally to take only the values 5 or 6, f_5 must equal 12. In addition, the formula puts no restrictions on f_6, and indeed a variety of fullerene molecules with a variety of numbers of hexagonal "faces" have been synthesized.

Group theory provides a methodological way of cataloging the vibrational symmetries of fullerenes, which can be linked to measurable energy spectra. There are 174 vibrational modes for a C_{60} molecule, but only 46 of these are potentially distinguishable. Group theoretic arguments based on irreducible representations and Schur's lemma pare this number dramatically, explaining why the observed infrared spectrum contains just four absorption lines. The same principles applied to the

---

[1]See, e.g., Chung and Sternberg (1993), on which this section is based.

scattering measured by Raman spectroscopy explain why that spectrum should have exactly 10 lines.

Graph theory is applied to fullerenes via Hückel theory, which forms the basis of an algorithm linking stability properties to the eigenvalues of a so-called adjacency matrix (representing, in this case, which pairs of carbons are bonded). Computation of the eigenvalues would be straightforward, at least for the smaller fullerenes, but more understanding comes about through mathematical analysis. Again by using Schur's lemma and other tools of group theory, the adjacency matrix can be decomposed into much smaller blocks that are amenable to providing clearer insight. For C_{60} this technique has led to closed-form solutions of certain matrices (Chung et al., 1993), which in turn suggest that C_{60} has a stability even greater than that of benzene.

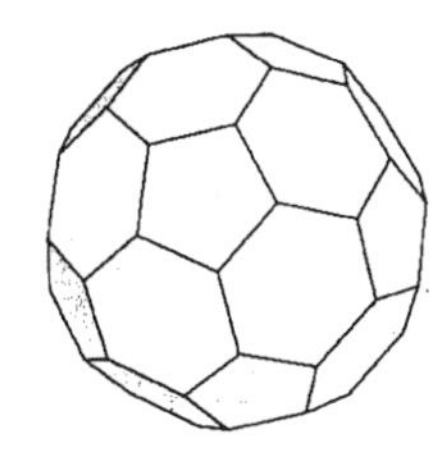

FIGURE 3.1 A buckyball.

References

Chung, F., and S. Sternberg, 1993, Mathematics and the buckyball, *American Scientist* 81:56–71.

Chung, F.R.K., D. Rockmore, and S. Sternberg, 1993, The symmetry and spectrum of the buckyball, 1993, preprint.

Quasicrystals

Interatomic and intermolecular forces have traditionally been central concerns in chemistry, not only because of the useful properties that they produce, but also because the crystallography that emerges from those forces when matter is solidified has been a dominant tool for structural analysis (see the section on X-ray crystallography below). Consequently, any striking deviation from conventional expectation about what interatomic or intermolecular forces can yield as solid-state ordering automatically concerns chemical researchers. The recently discovered quasi-crystalline state falls in this category.

That quasicrystals—orientationally ordered solids with local fivefold (or other classically unorthodox) symmetry but no spatial periodicity—really existed was demonstrated by Cahn et al. (1988) and Gratias et al. (1988). Their data were obtained from X-ray and neutron diffraction of $Al_{73}Mn_{21}Si_6$ single-phase icosahedral powder and were analyzed by Patterson analysis. Their work showed that the data were best considered as representing a "cut" of a periodic six-dimensional Patterson function. This work led to the definition of quasicrystals by the "cut-and-projection" method.

What is very interesting and perhaps not so well known is that Meyer (1972), 16 years earlier and motivated by problems in number theory, had built a mathematical structure that he called a "pseudolattice," which turns out to be the correct mathematical tool for the study of quasicrystals. Meyer's work establishes the basic harmonic analysis for quasicrystals In his book, he studied so-called Pisot and Salem numbers, which can be defined as follows: A Pisot number θ is a root of a polynomial with integer coefficients of degree m such that if $\theta_2, \ldots, \theta_m$ are the other roots, then $|\theta_i| < 1, i = 2, \ldots, m$. A Salem number is defined in the same way, but some of the inequalities are replaced by equalities.

One of the relations between Pisot or Salem numbers and quasicrystals is the following. Let $\Lambda \in \mathbb{R}^n$ be a quasicrystal. If $\theta > 1$ and $\theta\Lambda \subset \Lambda$, then θ is either a Pisot or a Salem number. Conversely, if θ is either a Pisot or a Salem number, then there exists a quasicrystal $\Lambda_\theta \subset \mathbb{R}^n$ such that $\theta\Lambda \subset \Lambda$.

In order to establish that result, Meyer had to establish a theory of harmonic analysis for quasicrystals.

Recently, Moody and Patera (1993) have used fairly sophisticated Lie group and Lie algebra ideas to study families of quasicrystals. In their work the mathematics has been used as a means of unifying various physical models of quasicrystals into one consistent picture.

References

Cahn, J.W., D. Gratias, and B. Mozer, 1988, Patterson Fourier analysis of the icosahedral (Al,Si)-Mn alloy, *Physical Review B* 38:1638–1642.

Gratias, D., J.W. Cahn, and B. Mozer, 1988, Six-dimensional Fourier analysis of the icosahedral $Al_{73}Mn_{21}Si_6$ alloy, *Physical Review B* 38:1643–1646.

Meyer, Y., 1972, *Algebraic Numbers and Harmonic Analysis*, North Holland Press. Amsterdam.

Moody, R.V., and J. Patera, 1993, Quasicrystals and icosians, *J. Phys. A. Math. Gen.* 26:2829.

Chemical Topology

Topology is a branch of mathematics that studies properties of objects that do not change when the object is elastically deformed. Topology allows stretching, shrinking, twisting—any sort of elastic deformation short of breaking and reassembling the object. The basic idea in topology is to relax the rigid Euclidean notion of congruence and replace it with more flexible notions of equivalence. A flexible molecule in solution does not maintain a fixed three-dimensional configuration. Such a molecule can assume a variety of configurations (referred to as "conformation" by chemists), driven from one to the other by thermal fluctuations, solvent effects, experimental manipulation, and so on. For small molecules with complicated molecular graphs, topology can aid in the prediction and detection of various types of spatial isomers (Walba, 1985; Simon, 1986; Walba et al., 1988). Recent triumphs in the chemical synthesis of molecules with novel topology include the molecular trefoil knot (Dietrich-Buchecker and Sauvage, 1989) (Figure 3.2) and the five interlocked rings of the self-assembling compound "olympiadane" (Amabilino et al., 1994). For larger molecules, given an initial topological state, one can identify all possible attainable configurations of the molecule and can detect when an agent (chemical or biological) has intervened to change the topological state (Wasserman and Cozzarelli, 1986).

Combinatorics, Graph Theory, and Chemical Isomer Enumeration

In the nineteenth century, Arthur Cayley produced a body of work involving the enumeration of certain types of trees (connected acyclic 1-complexes); some of the enumerated trees corresponded to the number of certain (combinatorially possible) chemical compounds (Cayley, 1857, 1877). This correspondence is obtained by abstracting a chemical molecule as a molecular graph; the vertices are the atoms, and the edges are the covalent bonds, with protons (hydrogen atoms) usually suppressed. Structural isomers correspond to the (abstract) isomorphism types of these graphs. Enumeration of isomorphism types uses group theory (permutation groups) to count the intrinsic (internal) symmetries of these graphs. This work was continued and expanded in the twentieth century by Polya (Polya and

Read, 1987) and others (Balaban, 1976).

A benefit to chemistry of combinatorial enumeration is that one can build a (mathematically) complete list of molecular graphs satisfying certain similarity (homology) parameters of chemical interest. As discussed in the section of this chapter beginning on page 37, the mathematical enumeration of possibilities is just the first in a series of steps that must be taken in order to extract chemically useful information (i.e., which of the molecular graphs are chemically realizable). Of those that are realizable, how does one store representative graphs in a database in a manner in which chemically relevant queries are feasible? The abstract (intrinsic) isomorphism type of a molecular graph carries little information about the three-dimensional (extrinsic) configuration, and hence the reactivity, of chemical realizations. How does one derive three-dimensional information from an abstract one-dimensional graph?

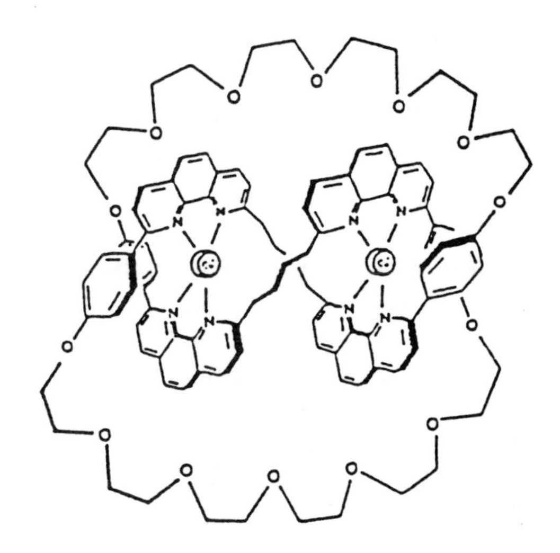

FIGURE 3.2 A synthetic trefoil. Reprinted, by permission, from Dietrich-Buchecker and Sauvage (1989). Copyright 1989 by VCH Publishers, Inc.

Analysis of Molecular Spectra by Using Cayley Trees

The high-resolution electronic spectra of polyatomic molecules are often very complex, consisting of many thousands of lines; yet they often contain significant information about molecular vibration/rotation structure and intramolecular dynamics. Traditionally this information has been extracted by assigning quantum numbers to lines based on an assumed zeroth-order Hamiltonian. However there are many molecules for which the choice of zeroth-order Hamiltonian is not clear, or for which the spectra are so strongly perturbed from a known zeroth-order Hamiltonian that it is not clear how or if a spectrum can be assigned. Statistical methods have been used to analyze these spectra, but most spectra contain information beyond the simple statistical limit. Another approach is to examine coarse-grained spectra (Gomez Llorente et al., 1989), as such spectra sometimes reveal hidden structure associated with short-time molecular dynamics that can easily be interpreted.

Recently there has been interest in using the methods of graph theory, in particular Cayley trees, to study complex spectra (Davis, 1993). The basic idea is to use the spectra to construct trees by coarse-graining the spectra over a hierarchy of time scales. An analysis of the statistical properties of these trees using methods taken from the cluster analysis literature (Gordon, 1987; Jain and Dubes, 1988) then provides a systematic way of locating hidden structure in the spectra. In addition, when quantum eigenstates are available for the spectra being studied, it is possible to use the trees to determine "smoothed" states that represent the underlying vibrational dynamics responsible for the hidden structure. This also provides a short time picture of the vibrational motions of the highly excited molecule, and in many cases it is possible to relate this to the underlying classical description of the vibrational motions, including features such as periodic orbits, bottlenecks to intramolecular vibrational redistribution, and Fermi resonances.

Group Theory, Topology, Geometry, and Stereochemistry

Stereochemistry studies the spatial configuration of molecules. To enumerate and distinguish stereoisomers, one must study symmetries of the molecule in 3-space (Cotton, 1971; Fujita, 1991).

Which of the intrinsic symmetries of the molecular graph are realizable by chemically reasonable spatial transformations? Of particular interest in this situation is the notion of chirality (see Figure 3.3). A molecule in space is chiral if it is *not* equivalent to the configuration obtained by a flexible transformation plus reflection in a plane.

Apart from energy minimization questions, the enumeration of spatial configurations and physical symmetries and chirality of relatively small molecules requires group theory, geometry, and topology. For small molecules, group theory is useful in distinguishing stereoisomers obtained by ligand substitution and in the study of dynamic symmetry of fluxional molecules (Longuet-Higgins, 1963; Smeyers, 1992). Group theory techniques include representation theory, group characters, and so on. Group theory has also been applied to quantum chemical problems such as symmetry-adapted functions for molecular orbital theory, ligand field theory, and molecular vibrations (Cotton, 1971; National Research Council, in preparation). By using group theory and representation theory, symmetry-invariant properties of physical interest can be studied (Kramer and Cin, 1980). Mulliken (1933), for example, used ideas from finite point groups and molecular orbital theory to assign "term symbols" (i.e., irreducible representation labels) to many excited states of small, highly symmetrical molecules.

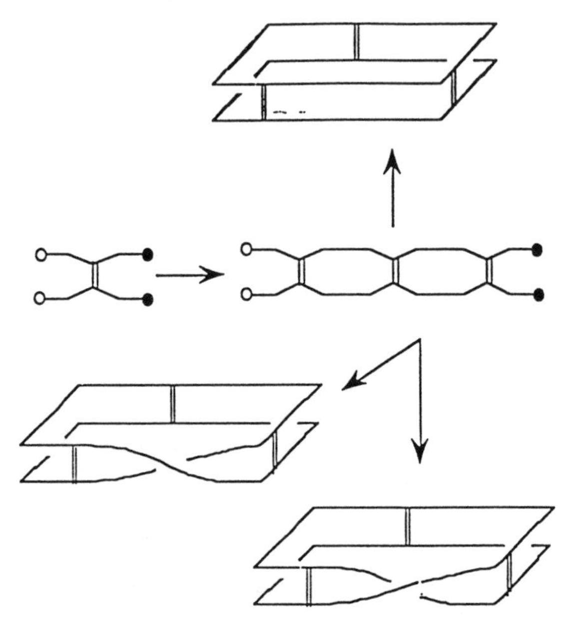

FIGURE 3.3 A chiral pair of synthetic Möbius molecules. Reprinted from Simon (1992) by permission of the American Mathematical Society.

New techniques for the efficient computation of the fast Fourier transform (FFT) for finite groups (Gordon, 1987; Diaconis and Rockmore, 1990) have potential applications in molecular spectroscopy and in understanding the symmetry of nonrigid molecules. The unitary group has been used in electronic structure theory to develop formulas for matrix elements, perturbation expansions, and coupled cluster expansions of the Hamiltonian written in second quantized form. For larger, more flexible molecules, the descriptive and computational ability of topology can be used to find topological barriers to the interconversion of a pair of spatial configurations or to the interconversion of protons in the molecule. Given the completely flexible equivalences of topology, in which energy, bond lengths, divalent vertices, and so on are disregarded, if two spatial molecular configurations (or specific atoms in the configuration) are topologically inequivalent, then they are physically inequivalent. Knowledge of the topological inequivalence of certain protons is useful in the analysis of NMR spectra (Walba, 1985; Walba et al., 1988). Topological considerations are sometimes very effective in detecting chirality of chemical compounds (Walba, 1985; Simon, 1986).

Topology of Polymers

Macromolecules such as synthetic polymers and biopolymers such as DNA, RNA, and proteins are very flexible and can exhibit high degrees of spatial complexity. Synthetic polymers can be very large, in some cases having on the order of 10^6 monomers in a polymer strand (e.g., polystyrene). Biopolymers (DNA, RNA, proteins) can also be extremely large, having hundreds to thousands of monomers. For example, in prokaryotes, DNA plasmids are on the order of 10^4 nucleotides, and bacterial chromosomes are on the order of 10^5 nucleotides. The effects of microscopic topological entanglement (knotting and linking) of polymer strands in a polymer melt are believed to be trapped by quenching (driving off the solvent, or cooling) and (in principle) will be observable in the macroscopic physical and chemical properties of the quenched polymer network (Edwards, 1968; Flory, 1976; deGennes, 1984).

Polymers in dilute solution can be modeled by means of self-avoiding walks on a lattice or piecewise-linear curves in 3-space. The lattice spacing serves to simulate volume exclusion; in the piecewise-linear case, volume exclusion can be modeled by assigning thickness to the linear segments of the chain. The degree of entanglement complexity of a polymer with itself (knotting) or with other polymers (linking) is believed to play a significant role in many physical and chemical processes, such as crystallization behavior and rheological properties (Edwards, 1968; Flory, 1976; deGennes, 1984). A linear polymer can be modeled as a self-avoiding walk (SAW) on the simple cubic lattice Z^3; a ring polymer can be modeled as a self-avoiding polygon (SAP) on Z^3. One can experimentally generate randomly embedded ring polymers by performing a cyclization (random closing) reaction on a dilute family of linear polymers of the same length N (Shaw and Wang, 1993; Rybenkov et al., 1993). A fundamental mathematical problem is to describe the yield of such a reaction: What is the distribution of knots and links produced by a random closing reaction, as a function of the length N and the concentration of linear substrate? A long-standing fundamental conjecture in this area was the Frisch-Wasserman-Delbruck (FWD) conjecture (Frisch and Wasserman, 1961; Delbruck, 1962): *The probability that a random polygon contains a knot tends to one as the number of edges tends to infinity.*

The FWD conjecture was recently solved with the development of a rigorous proof of the asymptotic inevitability of knotting (Soteros et al., 1992). The mathematical quantization of topological entanglement for short chains can be done by Monte Carlo simulation (Klenin et al., 1988). In Monte Carlo simulations, knotting and linking of random chains are computed in various models that include volume exclusion and some energetic terms; rigorous results in various models include the asymptotic inevitability of knotting in random chains and the (at least linear) growth of certain entanglement parameters with chain length. Monte Carlo simulation can also elucidate dynamic chemical phenomena such as electrophoresis (Slater and Noolandi, 1986) and adsorption (Smit and Siepmann, 1994). Recent striking advances in observation techniques such as fluorescence microscopy for single DNA molecules (Bustamante et al., 1990) make possible the verification and fine tuning of some of these mathematical theories of molecular conformation and dynamic molecular properties.

Knots, Links (Catenanes), and DNA

Mathematics and molecular biology continue to benefit from productive interaction, as described in the upcoming report *Calculating the Secrets of Life* (National Research Council, in press). One area of interaction is in the topology and geometry of DNA, because the spatial configuration of biopolymers is intimately related to function. For example, the DNA of all organisms has a complex and fascinating topology. Duplex DNA consists of a pair of DNA backbone strands (each strand is an alternating linear arrangement of sugar and phosphate moieties), and attached to each backbone are the nucleotides adenine, thymine, cytosine, and guanine. Adenine (A) binds only to thymine (T) by means of a double hydrogen bond, and cytosine (C) binds only to guanine (G) by means of a triple hydrogen bond; the bonded pairs A-T and C-G are called base pairs. The hydrogen bonds form the rungs of a ladder, and this ladder is twisted in space in the form of a right-hand helix (the usual Crick-Watson model for the primary structure of the double helix). In the double helix, one backbone strand winds around the other on the average of every 10.5 base pairs. The human genome is on the order of 3×10^9 base pairs, which amounts to some 3×10^8 interwindings. So, human chromosomal DNA can be viewed as two very long curves that are intertwined millions of times, linked to other curves, tied into knots, and subjected to four or five successive orders of coiling to convert it into a compact form for information storage. For information retrieval and cell viability, some geometric and topological features must be introduced, and others quickly removed. Some enzymes maintain the proper geometry and topology by passing one strand of DNA through another via an enzyme-bridged transient break in the DNA; these enzymes play crucial roles in cell metabolism, including segregation of daughter chromosomes at the termination of replication, and in maintaining proper in vivo (in the cell) DNA topology. Other enzymes break the DNA apart and recombine the ends by exchanging them. These enzymes regulate the expression of specific genes, mediate viral integration into and excision from the host genome, mediate transposition and repair of DNA, and generate antibody and genetic diversity. These enzymes perform incredible feats of topology at the molecular level; the description and quantization of spatial configuration and enzyme action require the language and computational machinery of geometry (White, 1992; Wolffe, 1994) and topology (Sumners, 1992).

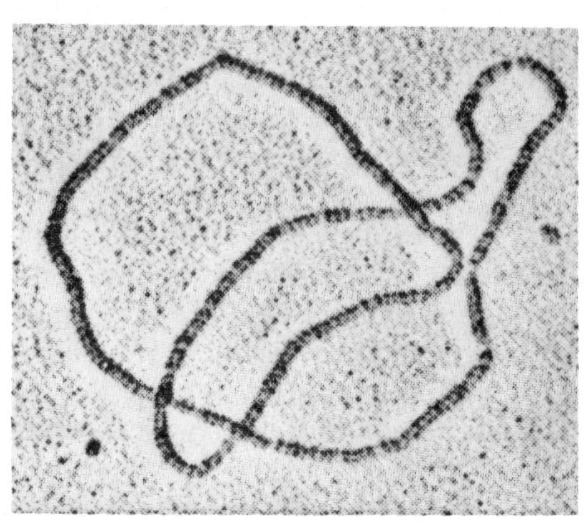

FIGURE 3.4 A DNA trefoil. Courtesy of N.R. Cozzarelli and A. Stasiak.

In the topological approach to enzymology (Wasserman and Cozzarelli, 1986), the topological invariance of knotted (see Figure 3.4) and catenated DNA during experimental work-up and the computational power of topology are exploited to capture information on enzyme action. For in vitro (in a test tube) experiments, an enzyme extracted from living cells is reacted with circular DNA substrate produced by cloning techniques. The enzyme reaction produces a topological signature in the form of an enzyme-specific family of supercoiled DNA knots and links (catenanes). By observing changes in DNA geometry (supercoiling, or interwinding of the DNA upon itself) and topology (knotting and linking) by means of gel electrophoresis and electron microscopy of the reaction products, the enzyme mechanism can be characterized mathematically (Sumners, 1992). Because of the enormous variety of knot and catenane structure, fine details of DNA structure and enzyme action can be selectively assayed.

References

Amabilino, D.B., P.R. Ashton, A.S. Reder, N. Spencer, and J.F. Stoddart, 1994, Olympiadane, *Angewandte Chemie (Int. Eng. Ed.)* 33:1286–1290.

Balaban, A.T., 1976, *Chemical Applications of Graph Theory*, Academic Press, London.

Bustamante, C., S. Gurrieri, and S.B. Smith, 1990, Observation of single DNA molecules during pulsed-field gel electrophoresis by fluorescence microscopy, *Methods: A Companion to Methods in Enzymology* 1:151–159.

Cayley, A., 1857, On the theory of analytic forms called trees, *Phil. Mag*. 13:172–176.

Cayley, A., 1877, On the number of univalent radicals C_nH_{2n+1}, *Phil. Mag*. Series 5, (3):34–35.

Cotton, F.A., 1971, *Chemical Applications of Group Theory,* Wiley-Interscience, New York.

Davis, M.J., 1993, Hierarchical analysis of molecular spectra, *J. Chem. Phys.* 98:2614.

deGennes, P.G., 1984, Tight knots, *Macromolecules* 17:703–704.

Delbruck, M., 1962, Knotting problems in biology, in *Mathematical Problems in the Biological Sciences* (Proceedings of Symposia in Applied Mathematics, vol.14), American Mathematical Society, Providence, R.I., pp. 55–63.

Diaconis, P., and D. Rockmore, 1990, Efficient computation of the Fourier transform for finite groups, *J. Am. Math. Soc.* 3:297–332.

Dietrich-Buchecker, C.O., and J.P. Sauvage, 1989, A synthetic molecular trefoil knot, *Angewandte Chemie (Int. Eng. Ed.)* 28:189–192.

Edwards, S.F., 1968, Statistical mechanics with topological constraints: II., *J. Phys. A.* 1:15–28.

Flory, P.J., 1976, Statistical thermodynamics of random networks, *Proc. R. Soc. Lond. A* 351:351–380.

Frisch, H.L., and E. Wasserman, 1961, Chemical topology, *J. Am. Chem. Soc.* 83:3789–3795.

Fujita, S., 1991, *Symmetry and Combinatorial Enumeration in Chemistry*, Springer-Verlag, Berlin.

Gomez Llorente, J.M., J. Zakrzewski, H.S. Taylor, and K.C. Kulander, 1989, Spectra in the chaotic region: Methods for extracting dynamic information, *J. Chem. Phys.* 90:1505–18.

Gordon, A.D., 1987, Parsimonious trees, *Classification* 4:85–101.

Jain, A.K., and R.C. Dubes, 1988, *Algorithms for Clustering Data*, Prentice Hall, Englewood Cliffs, New Jersey.

Klenin, K.V., A.V. Vologodskii, V.V. Anshlevich, A.M. Dykhne, and M.D. Frank-Kamenetskii, 1988, Effect of excluded volume on topological properties of circular DNA, *J. Biomol. Str. Dyn.* 5:1173–1185.

Kramer, P., and M.D. Cin, eds., 1980, *Groups, Systems and Many-Body Physics*, Vieweg and Sohn Verlagsgesellschaft mbH, Braunschweig.

Longuet-Higgins, H.C., 1963, The symmetry groups of non-rigid molecules, *Molec. Phys.* 6: 445–460.

Mulliken, R.S., 1933, Electronic structures of poly-atomic molecules and valence IV electronic states, quantum theory of the double bond, *Phys. Rev.* 43:279–302.

National Research Council, 1995, *Calculating the Secrets of Life*, National Academy Press, Washington, D.C., in press.

National Research Council, in preparation, *Group Theory: The Language of Symmetry in Science and Technology*, National Academy Press, Washington, D.C.

Polya, G., and R.C. Read, 1987, *Combinatorial Enumeration of Groups, Graphs, and Chemical Compounds*, Springer-Verlag, New York.

Rybenkov, V.V., N.R. Cozzarelli, and A.V. Vologodskii, 1993, Probability of DNA knotting and the effective diameter of the DNA double helix, *Proc. Nat. Acad. Sci. USA* 90:5307–5311.

Shaw, S.Y., and J.C. Wang, 1993, Knotting of a DNA chain during ring closure, *Science* 260: 533–536.

Simon, J., 1986, Topological chirality of certain molecules, *Topology* 25:229–235.

Simon, J., 1992, Knots and chemistry, in *New Scientific Applications of Geometry and Topology* (Proceedings of Symposia in Applied Mathematics, vol. 45), American Mathematical Society, Providence, R.I., pp. 97–130.

Slater, G.W., and J. Noolandi, 1986, On the reptation theory of gel electrophoresis, *Biopolymers* 25:431–454.

Smeyers, Y.G., 1992, Introduction to group theory for non-rigid molecules, *Adv. in Quantum Chemistry* 24.

Smit, B., and J.I. Siepmann, 1994, Simulating the adsorption of alkanes in zeolites, *Science* 264:1118–1120.

Soteros, C.E., D.W. Sumners, and S.G. Whittington, 1992, Entanglement complexity of graphs in Z^3, *Math. Proc. Camb. Phil. Soc.* 111:75–91.

Sumners, D.W., 1992, Knot theory and DNA, in *New Scientific Applications of Geometry and Topology* (Proceedings of Symposia in Applied Mathematics, vol. 45), American Mathematical Society, Providence, R.I., pp. 39–72.

Walba, D., 1985, Topological stereochemistry, *Tetrahedron* 41:3161–3212.

Walba, D.M., J. Simon, and F. Harary, 1988, Topicity of vertices and edges in the Mobius ladders; A topological result with chemical implications, *Tetrahedron Lett.* 29:731–734.

Wasserman, S.A., and N.R. Cozzarelli, 1986, Biochemical topology: Applications to DNA recombination and replication, *Science* 232:951–960.

White, J.H., 1992, Geometry and topology of DNA and DNA-protein interactions, in *New Scientific Applications of Geometry and Topology* (Proceedings of Symposia in Applied Mathematics, vol. 45), American Mathematical Society, Providence, R.I., pp. 17–37.

Wolffe, A.P., 1994, Architectural transcription factors, *Science* 264:1100–1101.

Graph Theory

Application of Graph Theory to Organizing Chemical Literature

To organize the chemistry literature for research or patent purposes, it is essential that scientists be able to search this literature by the chemical features of interest as well as with traditional text queries. Accordingly, a body of experience has been developed for using computers to recognize either total chemical structures or parts of them from an input structural diagram and to quickly identify in databases of millions of compounds those that match the search criteria (Ash et al., 1985). Currently there are several large comprehensive chemical databases such as those maintained by *Chemical Abstracts* (Dittman et al., 1983) and the Beilstein Institute. (The *Chemical Abstracts* substance database, for example, contains information on 13 million substances, including molecular formulas and structure diagrams where available.) Chemical and pharmaceutical companies maintain such chemical information databases of their own compounds using either commercial (e.g., MDLI,

DARC, Daylight) or self-written software.

A key element to the success of such chemical information systems was the recognition that a two-dimensional chemical structure diagram can be treated as a labeled graph (Sussenguth, 1965). Many algorithms and concepts from graph theory (Harary, 1976) have been applied to chemical informationproblems, for example, the concepts of graph isomorphism to identify whether a particular compound is in a database and subgraph isomorphism to identify compounds that contain substructures of interest, algorithms to detect the smallest set of smallest rings as an aid to unique atom numbering heuristics, and subgraph ("clique-detection"—see Box 3.4) algorithms to detect the maximum common substructure in two molecules. These ideas have recently been extended to provide rapid searches of databases of tens to hundreds of thousands of molecules to find those that match a three-dimensional query—typically based on distances, angles, and torsions between points in the stored three-dimensional structure of the molecules (Borman, 1989; Bures et al., 1994). Generally there are between 4 and 20 distance and angle constraints to be matched in a three-dimensional query: the number of hits decreases with the number of constraints. Recently, commercial programs have been updated to include consideration of conformational flexibility.

Graph algorithms are used increasingly to solve similar problems in molecular modeling and computational chemistry (Martin et al., 1992). For example, to use a molecular mechanics program to optimize a molecular structure, each atom must be assigned an atom type based on its substructural environment. Chemical information tools are used to recognize such substructures. Modeled three-dimensional structures are stored in a chemical information database with the result that it is easy to find a prebuilt analogue of a new compound one wishes to build. Others have devised programs that build three-dimensional structures of molecules from their two-dimensional structures by finding the maximum common overlapping substructures in a database of three-dimensional structures and piecing these together (Wipke and Hahn, 1988; Leach et al., 1990).

Methods based on graph theory have also been used to find common three-dimensional features within a set of molecules (Crandell and Smith, 1983; Brint and Willett, 1987; Martin et al., 1993). In particular, the Bron-Kerbosh clique-detection algorithm has been found to be especially fast (Brint and Willett, 1987). Such common three-dimensional features might represent a pharmacophore, the set of three-dimensional features that determines whether or not a molecule will show a particular biological activity. For example, Figure 3.6 depicts three different molecules that activate the D2 dopamine receptor. The figure shows that although these molecules look different in two dimensions, in three dimensions they share the arrangement of a hydrogen bond donor, its projection to a receptor hydrogen bond acceptor, a positively charged nitrogen, and its projection to an anionic site on the receptor (Martin et al., 1993).

Two-dimensional structures describe the connectedness of the atoms in a molecule. The training of a chemist involves learning how to translate these two-dimensional pictures into chemical properties. Thus, an OH means one thing to a chemist, but something different to other folks. Since molecules are really three-dimensional (with added dimensions of properties), it is important to translate the two-dimensional structure into three dimensions for computer processing. People have used the same type of graph-processing algorithms to detect parts of molecules that have certain three-dimensional properties and to then glue the three-dimensional pieces together much as when using a Tinkertoy. The methods are expert systems in that they use other knowledge, not first principles. They operate by using graph-matching ideas.

Clique-detection methods are also used to propose docking orientations of small molecules to macromolecules (Kuntz et al., 1982; Kuhl et al., 1984; Smellie et al., 1991; Kuntz, 1992). The computer program DOCK searches databases of tens of thousands of molecules to find those that fit into a macromolecular binding site of known three-dimensional structure (Kuntz, 1992). A number of structurally novel enzyme inhibitors have been identified by this means.

BOX 3.4 Clique Detection

In graph theory terms, a clique is a subgraph in which every node is connected to every other node. For three-dimensional molecular structures, the nodes of the graph might typically be the atoms of the structure labeled according to atomic symbol and the edges are the distances between the points. Clique-detection algorithms find cliques in an input graph that match a clique in a reference graph. That is, they find corresponding points in the two three-dimensional structures such that corresponding points are of the same type in the two structures and all corresponding interpoint distances are identical within some tolerance.

In Figure 3.5, two different cliques in the same molecule are indicated, as is one clique of a second molecule. As shown, the matching cliques in the two molecules can be superimposed. Notice that the points do not superimpose exactly since the lengths of the edges need only be identical within some tolerance.

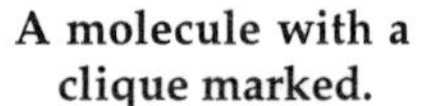

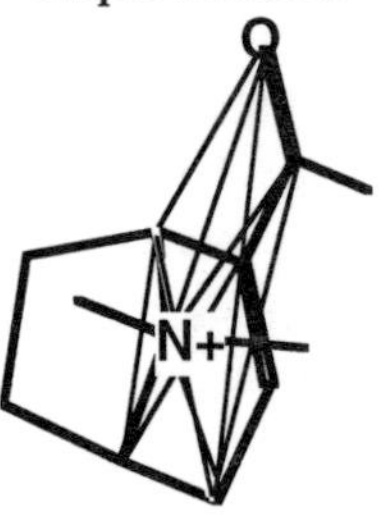

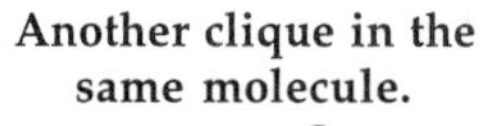

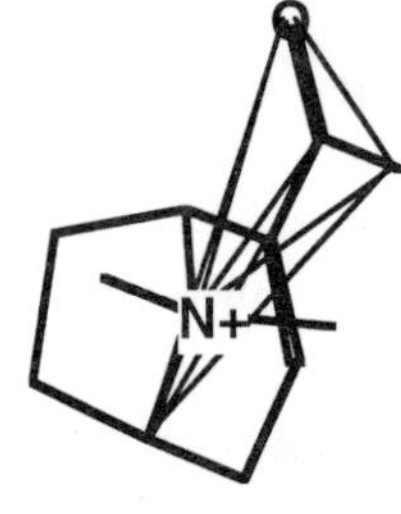

A second molecule with a clique marked.

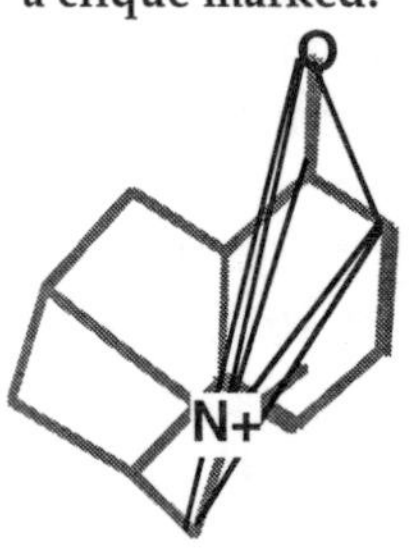

The two molecules superimposed over their matching cliques. The corresponding atoms are in shaded ellipsoids.

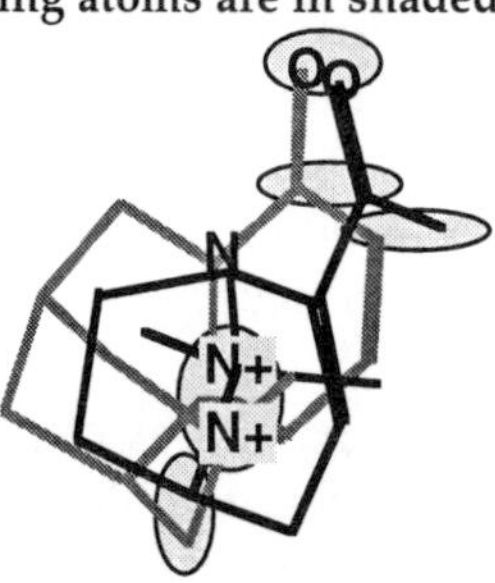

FIGURE 3.5 The basic operation of clique detection.

In general, these advances have not been the result of active collaborations between graph theorists and chemical information specialists. Rather, the chemical information specialists have followed the graph theory literature, and when a particular concept or algorithm seemed the appropriate solution for some problem, they would attempt to implement it. Sometimes this meant waiting until a graduate student with the particular skills was available. For example, in the case of the Ullman subgraph isomorphism algorithm (Ullman, 1976), Peter Willett suspected that it would be an improvement over the one used in chemical information systems at that time, and several graduate

2D chemical structures of three molecules that activate the D2 receptor.

3D overlay of these molecules in the proposed pharmacophore map.
The points for superposition are marked with an arrow.

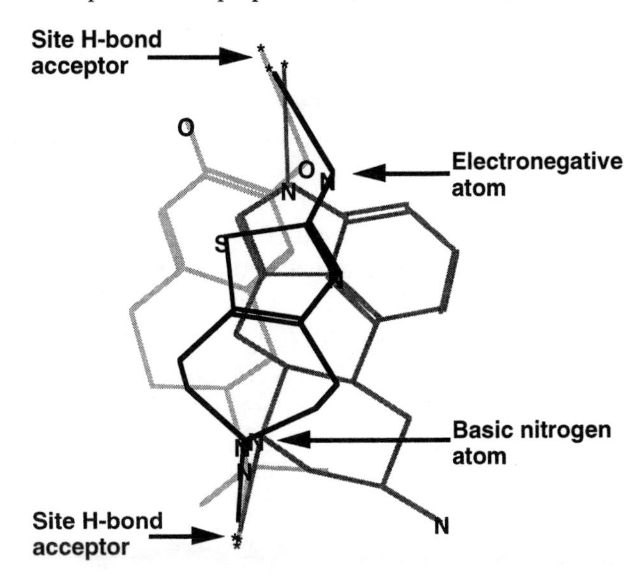

FIGURE 3.6 Development of a pharmacophore map. Reprinted (with adaptions) by permission from Martin et al. (1993). Copyright 1993 by ESCOM Science Publishers B.V.

students tried to implement it. It was not successfully implemented, however, until Andrew Brint, who was well trained in mathematics, took up the challenge. Since that time, 1986, the Ullman subgraph isomorphism algorithm has replaced the previous algorithms in all commercial chemical information systems (Willett, 1987).

Application of Graph Theory to Representation of Chemical Reactions

The synthesis of organic chemicals is an art that takes many years of training and experience to master. There are two aspects to synthesis: design of the synthetic pathway (what precursors and general reaction conditions will be used) and laboratory execution of the actual synthesis. Experienced synthetic chemists design the synthesis of new molecules based on their knowledge of the hundreds of types of reactions that can be done, their limitations in terms of the other structural features of the molecules that will survive the reaction conditions, and the usual success of the reactions in terms of side products and yield.

The complication with using the computer to aid in this process is that synthetic organic chemists rely primarily on information contained in two-dimensional chemical structure diagrams. A chemical reaction is simply a transformation of one diagram into another by combining and transforming starting materials into products. Hence, the expertise is organized intellectually in pictures, not numbers. The key insight that these structure diagrams can be described as labeled graphs enabled the use of the computer to process structure diagrams.

Computer programs for designing the synthesis of compounds also rely heavily on these algorithms (Wipke et al., 1978; Wipke and Rogers, 1984; Hanessian et al., 1990; Pensak and Corey, 1977; Johnson et al., 1992; Gasteiger et al., 1992). For example, it is essential to detect the smallest set of smallest rings since these form the basis of the synthetic strategy. Structure searching is used to ascertain if a proposed precursor is commercially available. Substructure searching identifies labile bonds to be broken in a retrosynthetic fashion. Maximal common subgraph algorithms have been used on the two-dimensional structures of products and reactants to perceive the part of a molecule unchanged in a chemical reaction (McGregor and Willett, 1980)—what is common between product and reactant is unchanged.

Two approaches have been used to apply the computer to the design of chemical syntheses. The first simply catalogues literature chemical reactions, transformations, with the associated starting materials and product(s), conditions, yield, and literature reference. For this application, the chemical reaction can be input as normally written and the computer can be used to detect which parts of the molecule are not changed. By definition, then, the parts of the molecular graph that change represent the chemical reaction. The unchanged parts represent the chemical context in which the reaction occurs. In graph theory terms, the unchanged parts are the maximum common subgraph of the reactants and products. The second type of computer program to aid the planning of synthesis involves the actual disconnection of the synthetic target into the proposed starting materials. Such computer programs treat molecules as labeled graphs but use rules encoded by experts to guide the proposed synthetic route. Because of the large number of rules to be encoded, such retrosynthetic programs are much less complete and of less general use than the reaction library programs.

References

Ash, J.E., S.E. Chubb, S.E. Ward, S.M. Welford, and P. Willett, 1985, *Communication, Storage and Retrieval of Chemical Information,* Ellis Horwood, Chichester.

Borman, S., 1989, Software adds new dimension to structure searching, *Chemical and Engineering News* 67:28-32.

Brint, A.T., and P. Willett, 1987, Algorithms for the identification of three-dimensional maximal common substructures, *J. Chemical Information and Computer Sciences* 27:152-158.

Bures, M.G., Y.C. Martin, and P. Willett, 1994, Searching techniques for databases of three-dimensional chemical structures, *Topics in Stereochemistry*, E.L. Eliel and S.H. Wilen, eds., John Wiley & Sons, New York, pp. 467-511.

Crandell, C.W., and D.H. Smith, 1983, Computer-assisted examination of compounds for common three-dimensional substructures, *J. Chemical Information and Computer Sciences* 23:186-197.

Dittmar, P.G., N.A. Farmer, W. Fisanick, R.C. Haines, and J. Mockus, 1983, The CAS ONLINE Search System. I. General system design and selection, generation and use of search screens, *J. Chemical Information and Computer Sciences* 23:93-102.

Gasteiger, J., W.-D. Ihlenfeldt, R. Fick, and J.R. Rose, 1992, Similarity concepts for the planning of organic reactions and syntheses, *J. Chemical Information and Computer Sciences* 32:700-712.

Hanessian, S., J. Franco, G. Gagnon, D. Laramee, and B. Laroche, 1990, Computer-assisted analysis and perception of stereochemical features in organic molecules using the CHIRON program, *J. Chemical Information and Computer Sciences* 30:413–425.

Harary, F., 1976, An exposition of graph theory, *Chemical Applications of Graph Theory*, A.T. Balaban ed., Academic Press, London, pp. 5–9.

Johnson, A.P., C. Marshall, and P.N. Judson, 1992, Starting material oriented retrosynthetic analysis in the LHASA program: 1. General description, *J. Chemical Information and Computer Sciences* 32:411–417.

Kuhl, F.S., G.M. Crippen, and D.K. Friesen, 1984, A combinatorial algorithm for calculating ligand binding, *J. Comput. Chem.* 5:24–34.

Kuntz, I.D., 1992, Structure-based strategies for drug design and discovery, *Science* 257:1078–1082.

Kuntz, I.D., J.M. Blaney, S.J. Oatley, R. Langridge, and T.J. Ferrin, 1982, A geometric approach to macromolecule-ligand interactions, *J. Mol. Biol.* 161:269–288.

Leach, A.R., D.P. Dolata, and K. Prout, 1990, Automated conformational analysis and structure generation: Algorithms for molecular perception, *J. Chemical Information and Computer Sciences* 30:13–16.

Martin, Y.C., 1992, 3D database searching in drug design, *J. Medicinal Chemistry* 35:2145–2154.

Martin, Y.C., K.-H. Kim, and M.G. Bures, 1992, Computer-assisted drug design in the 21st century, in *Medicinal Chemistry in the 21st Century* C.G. Wermuth, N. Koga, H. König, and B.W. Metcalf, eds., Blackwell, Cambridge, Mass., pp. 295–317.

Martin, Y.C., M.G. Bures, E.A. Danaher, J. DeLazzer, I. Lico, and P.A. Pavlik, 1993, A fast new approach to pharmacophore mapping and its application to dopaminergic and benzodiazepine agonists, *J. Computer-Aided Molecular Design* 7:83–102.

McGregor, J.J., and P. Willett, 1980, Use of a maximal common subgraph algorithm in the automatic identification of the ostensible bond changes occurring in chemical reactions, *J. Chemical Information and Computer Sciences* 21:137–140.

Pensak, D.A., and E.J. Corey, 1977, LHASA—logic and heuristics applied to synthetic analysis, *Computer Assisted Organic Synthesis*, ACS Symposium Series, Vol. 61, American Chemical Society, Washington D.C., pp. 1–32.

Smellie, A.S., G.M. Crippen, and W.G. Richards, 1991, Fast drug-receptor mapping by site-directed distances: A novel method of predicting new pharmacological leads, *J. Chemical Information and Computer Sciences* 31:386–392.

Sussenguth, E.H., 1965, A graph-theoretic algorithm for matching chemical structures, *J. Chemical Documentation* 5:36–43.

Ullman, J.R., 1976, An algorithm for subgraph isomorphism, *J. ACM* 16:31–42.

Willett, P., 1987, *Similarity and Clustering in Chemical Information Systems*, Research Studies Press, Letchworth.

Wipke, W.T., and D. Rogers, 1984, Artificial intelligence in organic synthesis, SST: Starting material selection strategies; An application of superstructure search, *J. Chemical Information and Computer Sciences* 24:71–81.

Wipke, W.T., and M.A. Hahn, 1988, AIMB: Analogy and intelligence in model building. System description and performance characteristics, *Tetrahedron Computer Methodology* 1:141.

Wipke, W., G.I. Ouchi, and S. Krishnan, 1978, Simulation and evaluation of chemical synthesis, SECS, *Artificial Intelligence* 11:173.

X-Ray Crystallography

An excellent source on the topic of X-ray crystallography is the article by Nobel laureate Herbert A. Hauptman (1990), from which the following is excerpted with permission of Plenum Publishing Corporation.

"It was recognized almost from the beginning that the diffraction pattern, the directions and intensities of the X-rays scattered by a crystal, is uniquely determined by the crystal structure; which is to say that if one knew the crystal structure—the arrangement of the atoms in the crystal—then one could calculate the diffraction pattern completely. It turns out that, conversely, diffraction patterns determine, in general, unique crystal and molecular structures, although this fact was not known until many years later. In short, the information content of a typical molecular structure coincides precisely with the information content of its diffraction pattern. It is a measure of the great advances made by the new science of X-ray crystallography that, nowadays, one can routinely transform the information content of a diffraction pattern into a molecular structure, at least for the so-called "small" molecules, that is those consisting of some 150 or fewer non-hydrogen atoms.

"Since X-rays, like ordinary visible light, are electromagnetic waves, they have a phase as well as an intensity, just as any other wave disturbance. In order to work backwards, from diffraction patterns to crystal and molecular structures, it turns out to be necessary to measure not only the intensities of the X-rays scattered by the crystal but their phases as well. However, the phases cannot be measured in the ordinary kind of diffraction experiment; they appear to be irretrievably lost. Only the intensities can be directly measured. This then gives rise to the central problem of X-ray crystallography, the so-called phase problem, how to deduce the values of the phases of the X-rays scattered by a crystal when only their intensities are known . . . .

"Because the needed phase information was lost in the diffraction experiment, it was thought that one could use arbitrary values for the phases associated with the measured intensities of the scattered X-rays. In this way one obtains a myriad of different crystal structures, all consistent with the known intensities. It therefore came to be generally believed that a procedure for going directly from the measured intensities to crystal structures could not, even in principle, be devised. By the same thinking, the problem of deducing the values of the individual phases from the diffraction intensities, the so-called phase problem, was also thought to be unsolvable, even in principle. It wasn't until the early 1950's, through the exploitation of special properties of molecular structures and through a simple mathematical argument, that these erroneous conclusions were finally refuted.

"My work on this problem started in 1948 about a year after I joined the Naval Research Laboratory in Washington, D.C. and initiated my collaboration with Jerome Karle. . . . The first important contribution that Karle and I made was the recognition that it would be necessary to exploit prior structural knowledge to transform the phase problem from an unsolvable one to one that was solvable, at least in principle. Our first step in this direction was to exploit the nonnegativity of the electron density function $p(r)$. Before our analysis was complete, however, David Harker and John Kasper [1948] published their famous paper . . . in which they derived inequalities in which the measured intensities restrict the possible values of the phases. This was a very mysterious paper, because nowhere in it does there appear any explicit mention of the basis for the inequality relations, and indeed the most important fact is conspicuous by its absence. It is simply that the electron density function is nonnegative everywhere. This fact is, however, implicit in Harker and Kasper's work. In very short order Karle and I completed our own analysis and derived the complete set of inequality relationships based on the nonnegativity of the electron density function [Karle and Hauptman, 1950] . . . . It includes the Harker-Kasper inequalities as a special case, and many others besides. Although the complete set of inequalities greatly restricts the values of the phases, the relations appear to be too intractable to be useful in applications, except for the simplest structures, and their potential has never been fully exploited . . . .

"Beyond any doubt our most important contribution during the early 1950's was the introduction of probabilistic techniques—in particular, use of the joint probability distribution of several diffraction intensities and the corresponding phases—as the central tool in the solution of the phase problem [Hauptman and Karle, 1953]. . . . We assumed to begin with that all positions of the atoms in the unit cell of the crystal were equally likely, or, in the language of mathematical probability, that the atomic position vectors were random variables, uniformly and independently distributed. With this assumption the intensities and phases of the scattered X-rays, as functions of the atomic position vectors, are also random variables, and one can use the methods of modern mathematical probability theory to calculate the joint probability distribution of any collection of intensities and phases. A suitably chosen joint probability distribution leads directly to the conditional probability distribution of a specified structure invariant, assuming again an appropriately chosen set of diffraction intensities. The conditional distribution in turn leads to the structure invariant, an estimate of which is given, for example, by its most probable value. Once one has a sufficiently large number of sufficiently reliable estimates of structure invariants, one can use standard techniques to calculate the values of the individual phases, provided that the process incorporates a recipe for specifying the origin.

"Although probabilistic methods played an essential role in the development of the direct method and provided it with its logical foundation, it must also be pointed out that nonprobabilistic methods also played an important part . . . . In particular the well known Sayre equation, a relationship of fundamental importance among measured magnitudes and unknown phases, continues to be useful to the present day and lies at the heart of some of the more successful computer programs for solving crystal structures.

"I cannot conclude this brief account of the early history of the direct methods of X-ray crystallography without also describing the reception this work received at the hands of the crystallographic community. This was, simply, extreme skepticism, if not outright hostility. . . .

"Today some 100,000 molecular structures are known, most determined by the direct methods, and about 5,000 new structures are added to the list every year. It is no exaggeration to say that modern structural chemistry owes its existence to this development . . . .

"Work on the phase problem continues to this day and applications to structures of ever increasing complexity continue to be made. It still appears that progress is made only in proportion to our ability to bring more powerful mathematical techniques to bear on this fascinating problem."

Remark

The committee can only add its belief that the last quoted sentence from Hauptman's account has wide applicability to problems of chemical interest, some of which are described in the next chapter.

References

Harker, D., and J.S. Kasper, 1948, *Acta Crystallogr.* 1:70.

Hauptman, H.A., 1990, History of x-ray crystallography, *Struct. Chem.* 6:617-620.

Hauptman, H., and J. Karle, 1953, *Solution of the Phase Problem I. The Centrosymmetric Crystal*, American Crystallographic Association Monograph No. 3, Polycrystal Book Service, Dayton, Ohio.

Karle, J., and H. Hauptman, 1950, *Acta Crystallogr.* 3:181.

4
MATHEMATICAL RESEARCH OPPORTUNITIES FROM THEORETICAL/COMPUTATIONAL CHEMISTRY

Introduction

This chapter highlights some of the most prominent research challenges from theoretical/computational chemistry that appear to be amenable to attack with the help of reasonable advances in the mathematical sciences. Most of the sections have as their starting point a challenge facing the chemical sciences, which is described in terms that should be accessible to the nonspecialist. The remaining sections—"Molecular Dynamics Algorithms," "Multivariate Minimization in Computational Chemistry," and "Fast Algebraic Transformation Methods"—contain discussions of topics from the mathematical sciences with specific insight into their relevance to theoretical/computational chemistry. Expositions and references are meant to give the mathematical reader sufficient insight and direction to be able into subsequently investigate the topic via deeper reading and especially by discussions with colleagues from the chemical sciences. Note that the topics surveyed in Chapter 3 are also sources of continuing research opportunity, some of which are identified therein.

As an overview of this chapter and a guide for navigating through it, the matrix in Figure 4.1 displays a subjective assessment of the depth of potential cross-fertilization between major challenges from theoretical and computational chemistry and relevant topics in the mathematical sciences. This matrix is based to some extent on intuition because it is an assessment of future research opportunities, not past results. An "H" in the matrix implies an overlap that appears clearly promising, while an "M" suggests that some synergy between the areas is likely. The absence of an H or an M should not be taken to imply that some clever person will not find an application of that technique to that problem at some point.

Two topics, polymers and protein folding, are consciously underrepresented in this report because the mathematical research opportunities related to these topics have been surveyed very recently by other reports from the National Research Council. For a discussion of mathematical opportunities related to the frontiers of polymer science, see pages 153–168 of *Polymer Science and Engineering: The Shifting Research Frontiers* (National Research Council, 1994). A chapter devoted to mathematical research into protein folding may be found in *Calculating the Secrets of Life* (National Research Council, 1995).

References

National Research Council, 1994, *Polymer Science and Engineering: The Shifting Research Frontiers*, National Academy Press, Washington, D.C.

National Research Council, 1995, *Calculating the Secrets of Life*, National Academy Press, Washington, D.C., in press.

Numerical Methods for Electronic Structure Theory

Most of the recent progress in theoretical chemistry has come from computation of numerical approximations to the solutions of realistic model problems. This has largely replaced earlier approaches based on analytic solutions to simplified model problems. Although there are interesting

	Quantum electronic structure	Molecular mechanics	Condensed-phase simulations	Density functionals	N-representability	Design of molecules	Construction of potential energy functions	Gas-phase dynamics	Polymers	Topography of potential energy surfaces	Biological macromolecules (including protein folding)
Adaptive and multiscale methods	H	M	M				M	M	M	M	M
Special bases	H		M	M				H			
Differential geometry		M				M	M		M	H	H
Functional analysis	H		M	H	H		H	M		M	M
Graph theory	M	M	M			H			M	M	M
Group theory	H		M		M	M		M	M		M
Optimization	H	H	H	H	M	H	H	M	H	H	H
Numerical linear algebra	H	H		M			M			M	
Number theory			M					M		M	
Pattern recognition	M	M	H			H		M	H	H	H
Probability and statistics	H		H		M	H		M	H	H	H
Several complex variables	H		H	M				M		M	M
Topology	M	M	H			H	M	M	H	H	H
Dynamical systems		M	H					H	M	H	M

FIGURE 4.1 Subjective assessment of depth of potential cross-fertilization between major areas of the mathematical sciences and theoretical and computational chemistry. An "H" implies an overlap that is clearly promising and an "M" suggests some synergy is likely.

intellectual problems involved with analytic properties of the exact solutions, progress to date with this approach has had little impact on chemistry.

In quantum chemistry, improved numerical approximations made possible by the greatly decreased cost (in dollars and personnel time) of computing have had an enormous impact. The *Journal of the American Chemical Society*, which once reluctantly accepted a limited number of articles reporting computational results, now has some computational aspect coupled with experimental results in more than half of the articles published.

The Schrödinger wave equation was introduced earlier, in Box 2.1. The primary method now used to obtain useful results may be described briefly as follows. The time-independent, nonrelativistic Schrödinger equation for a system of N electrons in the Coulomb field generated by K nuclei may be written in the Born-Oppenheimer approximation (using reduced units to eliminate mass and Planck's constant) as

$$H\psi = U\psi , \tag{1}$$

where

$$H = T + V \tag{2}$$

is the Hamiltonian operator with

$$T = -1/2 \sum_{i=1}^{N} \nabla_i^2 ,$$

$$V = -\sum_{i=1}^{N} \sum_{A=1}^{K} Z_A r_{iA}^{-1} + \sum_{i} \sum_{j>i} r_{ij}^{-1} , \tag{3}$$

$$\psi = \psi_k(r_1, r_2, \ldots, r_N; R_1, R_2, \ldots, R_K) ,$$

$$U = U_k(R_1, R_2, \ldots, R_K)$$

where r_i is the position vector for the ith electron and R_i is the position vector for the ith nucleus with charge Z_i .

This is a partial differential eigenvalue equation in $3N$ variables. The equation and its solutions are parameterized by the nuclear positions and charges. The eigenvalues U_k, viewed as functions of R_i, yield the potential energy surfaces for the molecule in its ground and excited states after including the repulsive internuclear interactions. The lowest eigenvalue, in particular, plays a fundamental role in understanding the chemical properties of a chemical system.

Equation (1) may be solved subject to either bound state or scattering boundary conditions. For a bound state, the function ψ should give finite values for the integrals $\int \psi^* \theta \psi d^3 r_1 d^3 r_2 \ldots$ for $\theta = 1$, T, or V, where the asterisk denotes complex conjugation. For a scattering state, the wavefunction may have an exp $(i\mathbf{k}\cdot\mathbf{r})$ behavior for large $|\mathbf{r}|$. The wavefunction is further subject to conditions of continuity everywhere and differentiability except at the singularities in V.

The Hamiltonian is symmetric under permutation of the electron coordinates. Consequently, the eigenfunctions can be chosen to form bases for irreducible representations of the permutation group. It has been found empirically that only a few irreducible representations actually correspond to physical reality and the rest are "excluded." This observation is summarized in the Pauli exclusion principle. In the standard labeling of the irreducible representations of the symmetric group by a partition of N into the sum of integers, only those partitions containing no integers greater than 2 are observed.

In quantum chemistry, this observation is usually implemented by introducing an additional discrete value, called the spin ξ_i, for each electron. This variable takes only the values $\pm 1/2$. The permutation operator is then defined to interchange the spin and position of a pair of electrons in the extended function $\psi(r_1\xi_1, r_2\xi_2, \ldots)$. The Pauli exclusion principle then is simply stated by observing that the only ψ that occur in nature change sign under every pairwise permutation of the electron coordinates.

Equation (1) has no closed-form solutions for nontrivial chemical systems. The most important problem in electronic structure theory is the rapid construction of useful approximate solutions to this equation. While chemists have made much progress on this problem, there is always the possibility that a fresh approach by mathematicians would lead to novel approaches. There is less possibility that changing the implementation of the methods now in use would lead to great improvement.

The dominant approach for obtaining approximate solutions at present is based on multiple expansions. First, a basis set is selected in three-dimensional space. Then this is transformed by a similarity transform to an orthonormal set of functions called "molecular orbitals." These orbitals are then used to construct Slater determinants, which are the simplest functions in $3N$ dimensions that obey the Pauli condition of antisymmetry under permutation of the positions and spins of the electrons. Finally, the wavefunction ψ is expanded as a linear combination of Slater determinants. Often, the transformation from basis functions to molecular orbitals is chosen to optimize an approximation to the wavefunction from a severely truncated expansion in Slater determinants. The coefficients in the Slater determinant expansion and the approximate eigenvalue U may be computed by the Rayleigh-Ritz variational principle, perturbation theory, cluster expansions, or some combination of those methods.

Slater determinants are determinants whose elements are n distinct orbitals for n distinct electrons.

In the **Hartree-Fock approximation** an exact wavefunction is replaced with an antisymmetrized product of single-particle orbitals (i.e., a Slater determinant).

All of these methods require formation of matrix elements of the Hamiltonian in the Slater determinant basis,

$$H_{IJ} = \sum_{\xi_1 \ldots \xi_N} \int \Phi_I \, H \, \Phi_J \, d^3\boldsymbol{r}_1 \, d^3\boldsymbol{r}_2 \ldots , \tag{4}$$

where Φ is a Slater determinant and the integration is over $3N$ dimensions. For a finite basis of orthonormal molecular orbitals, it is possible to replace this equation by a different one that gives the same matrix element

$$H_{IJ} = \langle \Phi_I | \tilde{H} | \Phi_J \rangle . \tag{5}$$

This is accomplished by utilizing an operator algebra that facilitates manipulation of Slater determinants, in which a_i is an annihilation operator and $a_i^\dagger$ is a creation operator. A correspondence can be established between the function space formed by all possible linear combinations of Slater determinants and the vector space formed by the creation operators so that

$$|\Phi_I\rangle = a_{i_N}^\dagger a_{i_{N-1}}^\dagger \ldots a_{i_1}^\dagger |0\rangle = |i_1 \ldots i_N\rangle \leftrightarrow \Phi_I . \tag{6}$$

That is, the Slater determinant, Φ_I, in which the orbital labels $i_1 \ldots i_N$ appear, corresponds to the abstract element $|\Phi_I\rangle$. The operator $\tilde{H}$, called the "second quantized Hamiltonian," takes the form

$$\tilde{H} = \sum h_{ij}\, a_i^{\dagger} a_j + 1/2 \sum g_{ijkl} a_j^{\dagger} a_i^{\dagger} a_k a_l \tag{7}$$

where, in terms of the orbitals ϕ_i,

$$h_{ij} = \int \phi_i^* (-1/2\nabla^2 - \sum_A Z_A r_A^{-1})\, \phi_j \mathrm{d}^3 r \quad \text{and}$$

$$g_{ijkl} = \int \phi_i^*(r_1) \phi_j^*(r_2) r_{12}^{-1} \phi_k(r_1) \phi_l(r_2)\, \mathrm{d}^3 r_1 \mathrm{d}^3 r_2 \, . \tag{8}$$

A major bottleneck in the present approach is the calculation of a large number of six-dimensional integrals g_{ijkl}. The choice of basis set is limited to those functions for which these integrals are readily computed.

The second major bottleneck is the actual calculation of U for this second quantized Hamiltonian. Although the values of $|U|$ for various chemical systems vary from 0 to several thousand, the range of U for interesting variations in the parameters for a given system is typically only 10^{-1} to 10^{-3}. Most programs determine U to a precision of 10^{-6}, regardless of the absolute magnitude. Because of the multiple expansions involved, the error in U is often 10^{-1} to 10^{-2}, but the error is often constant over the interesting range of nuclear position parameters to within $\pm 10\%$.

A major advantage of some of these procedures for obtaining U is that first and second derivatives with respect to the nuclear coordinate parameters can be obtained analytically. This has greatly facilitated searching U in this parameter space for minima, saddlepoints, and so on. The major disadvantage has been the very steep dependence of computer time on the number of electrons. Evaluating U for one set of parameters typically takes computing time proportional to N^k, with k ranging from 3 for the least accurate methods to 7 for the best methods now in common use. Finding one eigenvalue of $\tilde{H}$ in the full vector space is even more costly, with computer time proportional to M^{N+4}, where M is the number of basis functions. Consequently, accurate calculations are limited to about 40 electrons and an improvement of a factor of 10 in computer speed will not change this very much.

Although this approach is very productive, it is also limited to small chemical systems. Progress for somewhat larger systems can be made by use of the Hohenberg-Kohn and Kohn-Sham theorems to give a useful density functional theory. These assume that it is possible to find an effective one-body potential σ so that, by solving the one-electron Schrödinger equation

$$(-1/2\nabla^2 + \sigma)\phi_i = \epsilon_i \phi_i \, ,$$

the eigenfunctions ϕ_i will yield

$$\rho(r) = \sum |\phi_i(r)|^2 \, , \tag{9}$$

where ρ is the exact charge density of the system as defined by Equation (13) below, using the exact wavefunction defined by Equation (1). As before, σ, ϕ, and ϵ_i are parameterized by the nuclear charges and positions. Here σ is a functional of ρ. From ρ, the lowest potential energy surface U can be formed. The difficulty is that there is only an existence proof for σ and no systematic constructive procedure is yet known. Nevertheless, much progress has been made by choosing σ empirically so it will correctly reproduce the properties of simple model systems such as the uniform electron gas and free atoms. The attraction of this method is that the computational cost grows only as N^3. Effort is being made to improve this even further to obtain methods whose costs grow as N^2 or N.

Progress with density functional theory has been rapid in recent years. Direct solution of the Kohn-Sham equation on a three-dimensional grid now is possible, although basis set expansions are still more commonly used. The major limitation in this field is still the lack of a scheme for finding the effective potential that can guarantee the desired accuracy in U for all chemical systems of interest.

For even larger systems, solution of the Schrödinger equation seems hopeless. In this case, U is usually directly approximated by taking advantage of empirically observed near-transferability of parameters between similar chemical systems. This requires some input from the user of these programs to decide which atoms are bonded. Then, near a local minimum in U, it is possible to approximate U as a sum,

$$U \simeq \sum \Delta U \text{ (for displacements from normal bond lengths and bond angles)}$$

$$+ \sum \Delta U \text{ (for pairwise nonbonded interactions).} \tag{10}$$

Each pair of nonbonded atoms can be assigned parameters transferred from experimental data to allow a calculation of a $\Delta U(R_{AB})$ energy contribution. Similarly, each chemically distinct type of bond can be assigned an energy for displacement from its usual bond length. Functions constructed in this way have accuracy approaching the desired 10^{-3} level needed for chemical prediction. This approach is now used for widely different problems, such as rational drug design, structure of liquids, predicting the shape of moderate-sized organic molecules, and protein folding.

There are many problems in numerical analysis and data handling associated with the present methods. These include the generation and manipulation of large numbers of six-dimensional integrals, finding eigenvalues and eigenvectors of large matrices, and searching a complicated function for global and local minima and saddle points. There are also important questions about the construction of optimum expansion functions for most rapid convergence. At present, there are no useful error bounds for the energy or other properties derived from the wavefunction.

The N- and V-Representability Problems

Conventional approaches for computing the solution to the wavefunction have a strong dependence on the number of electrons. Therefore, searches are constantly under way for methods of comparable accuracy with a better scaling. In this regard, people have considered methods based on density matrices, density functionals, Monte Carlo diffusion equations, effective core potentials, and so on. In particular, because the energy of an atom or molecule is a linear function of the density matrix and the one- and two-body distribution functions derived from it, density matrix methods raise the hope that one could dispense with computing the associated $3N$-dimensional wavefunction and deal with simpler three-dimensional density functions. A number of mathematical issues are raised by the attempt to reformulate computational chemistry in terms of particle distribution functions instead of wavefunctions.

The N-body distribution function is given simply as the product of the wavefunction and its complex conjugate,

$$\Gamma_i^{(N)}(\boldsymbol{x}|\boldsymbol{x}') = \psi_i(\boldsymbol{x})\psi_i^*(\boldsymbol{x}'). \tag{11}$$

Here, $\boldsymbol{x}$ symbolizes the collection of coordinates $\boldsymbol{x}_1, \boldsymbol{x}_2, \ldots, \boldsymbol{x}_N$, describing the positions $\boldsymbol{r}_i$ and spins ξ_i of all N particles. In the chemical literature, this quantity is usually called the N-body density matrix. This may be averaged over an ensemble with an arbitrary set of probabilities (weights) w_i to give the ensemble distribution,

$$\Gamma^{(N)} = \Sigma w_i \Gamma_i^{(N)}, \qquad w_i \geq 0, \quad \Sigma w_i = 1. \tag{12}$$

Because of permutational symmetry, all identical particles enter this distribution in an equivalent way. From the N-body distribution, one- and two-body position distribution functions can be obtained by integration over the other coordinates:

$$\begin{aligned}
\Gamma^{(2)} &= \sum_{\zeta_3 \ldots \zeta_N} \int \cdots \int \mathrm{d}^3 r_3 \ldots \mathrm{d}^3 r_N \, \Gamma^{(N)}(x_1 x_2 x_3 \ldots x_N \,|\, x_1' x_2' x_3 \ldots x_N), \\
P(r_1, r_2) &= \sum_{\zeta_1 \zeta_2} \Gamma^{(2)}(x_1 x_2 | x_1 x_2), \\
\gamma(r_{12}) &= \int \mathrm{d}^3 r_2 P(r_2 + r_{12}, r_2), \\
\rho(r_1) &= \int \mathrm{d}^3 r_2 P(r_1, r_2).
\end{aligned} \tag{13}$$

The distribution function $P(x_1,x_2)$ is also called the two-body reduced density matrix. Similarly, the density matrix may first be Fourier transformed and then used to derive the one-body momentum distribution

$$\pi(p_1) = (2\pi)^{-3} \sum_{\zeta_1' \zeta_1 \zeta_2} \int\int\int \mathrm{d}^3 r_2 \mathrm{d}^3 r_1 \mathrm{d}^3 r_1' \, e^{-ip_1 r_1} \, e^{+ip_1 r_1'} \, \Gamma^{(2)}(x_1 x_2 \,|\, x_1' x_2) \, \delta_{\zeta_1 \zeta_1'} . \tag{14}$$

For most cases of interest, knowledge of the one- and two-body position distribution functions and the one-body momentum distribution function suffices to determine the energy. For the Hamiltonian H with given external potential $V_{ex}(r_i)$ and given two-body interaction potentials $g(r_{ij})$,

$$H = \sum_{i=1}^{N} \frac{p_i^2}{2m} + \sum_i V_{ex}(r_i) + \sum_{i>j} g(r_{ij}), \tag{15}$$

the energy is given by

$$\begin{aligned}
E &= N \int \frac{p^2}{2m} \pi(p) \mathrm{d}^3 p + N \int V_{ex}(r) \rho(r) \mathrm{d}^3 r \\
&+ \frac{N(N-1)}{2} \int g(r_{12}) \, \gamma(r_{12}) \mathrm{d}^3 r_{12} .
\end{aligned} \tag{16}$$

There are a number of outstanding mathematical problems associated with these distributions. For many potentials, the behavior of $\rho(r)$ is known near the singularities of the potential. Similarly, the behavior of $\gamma(r_{12})$ is known for a Coulomb interaction near the singularities, and the form of $\pi(p)$ is known for large and small p. The N-representability problem consists of finding the conditions on this set of three functions such that they could all come from the same N-body density matrix. Many inequalities are known, but no general solution has been obtained. The V-representability problem consists of determining the further restriction imposed by considering only those distributions that can come from a wavefunction that is the eigenfunction of some H for a fixed g.

The two-body reduced density matrix $\Gamma^{(2)}$ can be integrated to give the one-body reduced density matrix

$$\Gamma^{(1)}(x_1 \mid x_1') = \int d^3x_2\, \Gamma^{(2)}(x_1 x_2 \mid x_1' x_2). \tag{17}$$

Density matrices may be treated as integral kernels and factored in terms of their eigenfunctions

$$\Gamma^{(1)}(x_1 \mid x_1') = \Sigma\, \eta_i f_i(x_1) f_1^*(x_1') \tag{18}$$

where

$$\int \Gamma^{(1)}(x_1 \mid x_1') f_i(x_1') d^3x_1' = \eta_i f_i(x_1). \tag{19}$$

These functions f_i are known as natural orbitals. It is conjectured, but not proved, that they form a complete set when $\Gamma^{(1)}$ is derived from an exact eigenfunction of H with Coulomb interactions. The complementary functions

$$F_i(x_1 \ldots x_{N-1}) = \int d^3x_N\, f_i(x_N)\, \psi(x_1 \ldots x_N) \tag{20}$$

are certainly not complete in the $3(N - 1)$ dimensional space. The extended Koopmans theorem claims, however, that if ψ is the exact eigenfunction of H for an N-electron system, then the ground state wavefunction of the $N - 1$ electron system may be expanded exactly in the set of F_i. Both "proofs" and "disproofs" of this conjecture exist in the literature (see Morrison, 1993, and Sundholm and Olsen, 1993, for a recent exchange of opinions and a list of relevant earlier papers).

Because the energy is a linear function of the one- and two-body distribution functions, the variational minimum will lie on the boundary determined by the N-representability conditions. Unfortunately, only an incomplete set of necessary conditions are known, but these are already so complex that further work in this area has been abandoned by chemists.

In density functional theory, only the density $\rho(r)$ is used. Hohenberg and Kohn showed that the energy is a functional of this density for a fixed two-body potential $g(r_{ij})$. Density functional theory has become of practical importance, but progress is hindered by lack of knowledge of the properties of the functional and lack of a systematic procedure for constructing a convergent sequence of functionals. Some past work in this field has been summarized in several monographs (Davidson, 1976; Dahl and Avery, 1984; Kryacho and Ludena, 1989; March, 1989; Parr and Yang, 1989; Sham and Schlinter, 1989; Gadre and Pathah, 1991).

References

Dahl, J.P., and J. Avery, 1984, *Local Density Approximations in Quantum Chemistry and Solid State Physics*, Plenum Press, New York.

Davidson, E.R., 1976, *Reduced Density Matrices in Quantum Chemistry*, Academic Press, New York.

Gadre and Pathah, 1991, *Advances in Quantum Chemistry* 22:211.

Kryacho, E.S., and E.V. Ludena, 1989, *Density Functional Theory of Many-Electron Systems*, Kluwer Press, Dordrecht.

March, N.H., 1989, *Electron Density Theory of Atoms and Molecules*, Academic Press, New York.

Morrison, R.C., 1993, The exactness of the extended Koopmans Theorem—A numerical study—Comment, *J. Chem. Phys.* 99:6221.

Parr, R.G., and W. Yang, 1989, *Density Functional Theory of Atoms and Molecules*, Oxford University Press,

Oxford.

Sham, L.J., and M. Schlinter, 1989, *Principles and Applications of Density Functional Theory*, World Scientific, Teaneck, N.J.

Sundholm, D., and J. Olsen, 1993, The exactness of the extended Koopmans theorem—A numerical study—Response, *J. Chem. Phys.* 99:6222.

Melding of Quantum Mechanics with Simpler Models

A daunting challenge for the future is to accurately model chemical reactions in phases and at the active site of enzymes. An ability to do so would be of great importance in designing new biological catalysts as well as fully understanding the chemical mechanism of those that already exist. This would be of significant technological as well as scientific importance. One could imagine that many new molecules could be made and made much more efficiently by such catalysts.

Methods to do this in an approximate way have been available since 1976 (Warshel and Levitt, 1976) and have involved using the Schrödinger equation (Equation 1 or a variant or empiricized form of it) for the parts of the system where bonds are being made or broken and thus the electronic structure is changing, combined with representations such as Equation (10), which assume transferable electronic structure, for the remaining atoms of the system. Typically, the number of atoms for which Equation (1) must be solved is much smaller, of the order of 20 to 30, than the number in the whole system, which for the chemical reactions mentioned is typically more than a thousand.

In fact, some exciting results for simple reactions involving organic molecules in different solvents have been achieved (Blake and Jorgensen, 1991). In these cases, one has solved Equation (1) to high accuracy for a simple reactive pathway in vacuo and then, employing these energies, has used free energy calculation methods to evaluate the solvation free energy of different structures along the reactive pathway. This is in some sense a proof of concept for the combined application of Equations (1) and (10) because impressive agreement with experiment has been achieved in these simple, well-defined cases.

For more complex cases, such as enzyme reactions, the reaction pathway might involve many steps, and some of the reacting groups are chemically bonded to the protein, thus requiring some additional technical challenges in simulating the atoms at the junction between those that are participating in the chemical reactions and those that are not. In addition, one might have to consider many conformations of the enzyme and its substrate and accurately represent their relative energy by using the energy function of Equation (10), all the while considering the electronic energy (Equation 1) and the relative total free energy of the system.

As noted above, progress on this problem has been made when employing much simplified representations of the electronic structure of the system, which enable the solution of equations such as Equation (1) for the few "quantum mechanical" atoms as rapidly or more so than the classical molecular dynamical equations of motion, using Equation (10) as a potential energy (Field et al., 1990; Warshel, 1991).

These methods use semi-empirical or empirical valence bond approximations to solve Equation (1). Although these methods are not highly accurate, the use of non-empirical quantum mechanical

BOX 4.1 Electronic Phase Transitions

The electronic properties of extended bulk materials such as metals and superconductors have been studied extensively by physicists. However, as the field has progressed from simple elemental materials to more complex synthetic ones, computational chemistry has come to the fore in addressing these problems. The "exotic materials" include the organic superconductors such as Bechgaard salts, polyacetylene, the fullerenes, and the high-temperature superconductors based on copper oxides. In addition, many new amorphous materials are technologically interesting in their electronic properties. For instance, electronic conduction in polymers, such as polyvinylcarbazole, is essential to some xerographic processes. The chemical complexity of these systems puts a premium on understanding the fundamental physics in new ways that do not usually rely on the simple symmetries present in elemental materials. All of these systems exhibit electronic phase transitions as the chemical composition or doping is changed: their electronic states change qualitatively. Sometimes this transition is from being an insulator to a metal, sometimes from a metal to a superconductor or to some complex magnetically ordered structure.

One of the simplest electronic phase transitions is the transition between extended and localized states of a single electron moving in a random potential. Even though this problem is at the heart of the study of the electronic properties of any disordered material, the traditional methods of simply combining rough semiquantitative theories and experiment have been insufficiently powerful to resolve all of the important issues. One reason is that real materials have many physical influences in addition to disorder, such as interactions of vibrations and interactions between different electrons in the same material. Some mathematical work, such as rigorous theorems related to one-dimensional Anderson localization, has already helped in understanding this problem. On the other hand, loopholes in these theorems have been uncovered when the disorder is of a special, correlated kind. For instance, certain 1-D systems do not have only localized states as the simple theorems had indicated. These unusual sorts of extended states arise in systems with certain kinds of correlated disorder or with quasi-periodic Hamiltonians.

Interestingly, far from being a mathematical curiosity, these exceptions to the simple theorems about one-dimensional localization seem to be at the heart of understanding the behavior of materials such as polyacetylene. In two- or three-dimensional materials, experiment has amply demonstrated the existence of both extended and localized states. However, there are still numerous controversies about the applicability of simple phase transition ideas to these electronic phase transitions. Are they described by the usual scaling phenomenology of ordinary thermal phase transitions?

It has been argued that, in fact, such descriptions may break down because the wavefunctions at the transition are multifractal. Thus, the study of these electronic phase transitions has much in common with problems addressed in quantum chaos, where the structure of the wavefunctions needs to be understood in a statistical way. The interacting electron systems and their phase transitions also carry mathematical challenges. In one dimension, the interactions between electrons again can cause them to behave as if they were insulating. These one-dimensional many-electron problems lead to exactly solvable Schrödinger equations. The exact integrability of these classical models is connected with conformal invariance and the existence of solitons of nonlinear partial differential equations in one dimension. The question of whether such electronic interaction effects give rise to new phases for higher-dimensional systems doubtless has connections with the problem of solitons and exact integrable systems in higher dimensions—a problem that has attracted many in the area of partial differential equations.

Finally, a large number of interesting phase transitions occur in disordered systems that also have interactions. These include the Kondo effect in which isolated electronic sites behave as if they have spins that can be quenched, as well as the exotic spin glass phases that have proved useful at least as analogies in many other areas of chemistry and biology. At the moment, the most useful theory of these system is based largely on the use of the unrestricted Hartree-Fock approximation. There are numerous mathematical questions connected with the Hartree-Fock problem for such disordered systems. Many of the ideas invented by Lieb in his proof of the stability of matter may have practical use here. Again, the rigorous mathematical analysis can be of significant help in understanding whether or not certain approximations can be used confidently in elucidating the qualitative physics. In addition, the same issues will arise when quantitative calculations are contemplated for specific materials.

The study of electronic phase transitions in clusters will make these issues even more pronounced. The usual theory of phase transitions concentrates on infinite systems, and only dominant thermodynamic contributions are computed. For clusters, the finite size effects and the asymptotics of the approach to the infinite limit become important mathematical problems.

methods for systems of 20 to 30 atoms (the traditional ab initio approach) requires 10^4 to 10^6 more computations than the semi-empirical or empirical approaches and much more time than it takes to solve the classical equations of motion for the rest of the system (Singh and Kollman, 1986).

To elaborate, calculation of the free energy of a complex chemical system by using classical molecular dynamics (Kollman, 1993) requires one to calculate the energy of the system and its gradient with respect to all the $3N$ coordinates. This can be done for noncovalent processes (those using only Equation 10) quite efficiently because the energy function in Equation (10) is very simple and its derivatives are quick and easy to evaluate. When one adds quantum mechanical (bond making or breaking) effects via Equation (1), in order to make the calculation of the free energy tractable, one must be able to evaluate the quantum mechanical energy and its gradient for the few quantum mechanical atoms as rapidly as the classical molecular mechanical energy and gradient for the thousands of atoms in the remainder of the system. This can be done by using simpler empirical (Warshel, 1991) and, to a reasonable approximation, semi-empirical (Field et al., 1990) quantum mechanical methods, but not with the first principle ab initio methods.

Density functional methods (Labanowski and Andzelm, 1991), particularly the divide-and-conquer strategy (Yang, 1991), show promise in leading to accurate and rapid solutions of Equation (9) for the electronic structure, but they are still a long way from being fully developed, so one cannot tell how efficient and useful they will be in this regard.

Thus, accurate simulation of chemical reactions at the active sites of macromolecules will likely require significant progress in the conformational search problem, even if one considers only the active site of the enzyme. As should be emphasized, the "conformational search problem" requires one not only to consider many conformations, but also to rank their relative free energy in solution. On top of this, one places the problem of accurate and very rapid electronic structure calculations. The above problems are very challenging conceptually, practically, and computationally.

References

Blake, J.F., and W. Jorgensen, 1991, Solvent effects on a Diels-Alder reaction from computer simulations, *J. Am. Chem. Soc.* 113:7430–7432.

Field, M.J., P.A. Bash, and M. Karplus, 1990, A combined quantum mechanical and molecular mechanical potential for molecular dynamics simulations, *J. Comput. Chem.* 11:700–733.

Kollman, P., 1993, Free energy Calculations—Applications to chemical and biochemical phenomena, *Chem. Rev.* 93:2395–2417.

Labanowski, J., and J.W. Andzelm, eds., 1991, *Density Functional Methods in Chemistry*, Springer Verlag, New York.

Singh, U.C., and P.A. Kollman, 1986, A combined ab initio QM/MM method for carrying out simulations on complex systems: Application to the CH_3Cl + Cl^- exchange reaction and gas phase protonation of polyethers, *J. Comput. Chem.* 7:718–730.

Warshel, A., 1991, *Computer Modeling of Chemical Reactions in Enzymes and Solutions*, John Wiley, New York.

Warshel, A., and M. Levitt, 1976, Theoretical studies of enzymic reactions: Dielectric, electrostatic and steric stabilization of the carbonium ion in the reaction of lysozyme, *J. Mol. Biol.* 103:227–249.

Yang, W., 1991, Direct calculation of electron density in density-functional theory–implementation for benzene and a tetrapeptide, *Phys. Rev. A.* 44:7823–7826.

Enhanced Sampling

The multiple minima problem, discussed elsewhere in this report, can be addressed by approaches other than optimization. Specifically, reasonable sampling strategies can be devised to scan configuration space in the hope of obtaining information about many local minima, maxima, and saddlepoints. Moreover, these strategies can incorporate a variety of chemical information, such as interproton distances from NMR, van der Waals radii (for hard-core exclusion), and secondary-structure elements. Useful search strategies today involve Monte Carlo, molecular dynamics (MD), Brownian and Langevin dynamics, calculations of free-energy perturbations, high-temperature simulations, normal-mode analyses, and various enhanced sampling techniques that involve hybrids of all of the above. As reflected by those many approaches, the problem of inadequate sampling of configurational space is receiving increasing attention as a realization emerges that faster and more powerful computers alone cannot solve this problem in the near future. New methodologies and a hierarchy of approaches at different levels of resolution—in combination with experiment—are needed to attack this sampling problem to advance current capabilities of computational chemistry in connection with biomolecules.

A specific problem involves numerical integration in the context of MD simulations. In this technique, molecular motion is propagated by numerically integrating the classical equations of motion governing a molecular system under the influence of a specified force field (McCammon and Harvey, 1987; Allen and Tildesley, 1990). In theory, MD simulations can provide extensive spatial and temporal information. However, inadequate sampling limits the scope of the results that can be obtained in practice. Similar issues arise in other chemical applications, such as quantum mechanics, and the development of improved integration schemes will advance the systems and types of processes that can be simulated on modern computers.

Numerical Methods for Solving Ordinary Differential Equations

Many problems in chemistry can be reduced to the solution of systems of coupled ordinary differential equations (ODEs). Examples include classical and Langevin dynamics, rate equations of kinetic theory, and the time-dependent Schrödinger equation when expanded in a basis set. Thus, numerical integrators used to solve these equations are fundamental tools in computational/theoretical chemistry, and any significant improvement in these integrators (e.g., speedup, long-time stability) results in advances throughout the field.

The technology of numerical integrators for solving ODEs has a long history with significant interplay among mathematics, physics, and chemistry. Many of the earliest integrators, such as Runge-Kutta and predictor-corrector integrators, are still in common use, but there have also been recent advances, driven in part by the need for methods that can treat multiple time scales and have greater stability for large-scale coupled nonlinear oscillators commonly found in MD of polymers and biological macromolecules. The long-time stability of integrators for such systems is a challenging area of mathematical analysis research; perhaps the chemical applications described here will stimulate important developments.

Symplectic Integrators

Symplectic integrators have recently gained attention in the mathematical community and were quickly adapted for use in dynamics calculations in chemistry because of their favorable properties. In applications to Hamiltonian systems, symplectic integrators have the property of building in Liouville's theorem, whereby areas in phase space are preserved as the system evolves in time. This strong conservation property translates into stability over long-time integrations, an important property in MD calculations involving millions and more steps. One consequence of this for constant-energy MD simulations is that except for fluctuations, symplectic integrators with small time steps conserve energy for very long times, whereas nonsymplectic integrators typically introduce a systematic drift in the total energy. Time reversibility is another useful practical property of symplectic integrators.

A **symplectic integrator** is any of a class of numerical algorithms for the integration of classical many-body equations of motion that possess favorable properties such as area preservation, energy conservation, and time reversibility.

Symplecticness may also be used in determining numerical solutions to the Schrödinger equation. There is an equivalent representation of quantum mechanics in terms of Hamilton's equations (Gray and Verosky, 1994) that makes possible the use of integrators for the quantum dynamics studies that are used for classical dynamics. Area-preserving mapping are also of interest in their own right in studies of dynamical systems (Meiss, 1992).

Symplectic integrators may be *implicit* or *explicit*. In explicit methods, the solution at the end of the time step is obtained by performing operations on the variables at the beginning of each time step. Symbolically, we write $y^{n+1} = f(y^n, \Delta t, \ldots)$, where f is some nonlinear function, Δt is the time step, y^n is the approximation to the solution y at time $n\Delta t$, and the dots indicate other parameters or previous solutions (e.g., y^{n-1}, y^{n-2}). With implicit integrators, the final solutions are functions of both the initial and the final variables ($y^{n+1} = f(y^{n+1}, y^n, \Delta t, \ldots)$), and so coupled nonlinear equations must generally be solved at each time step to propagate the trajectory. The explicit versions generally involve simple algorithms that (for propagation only) use modest memory, while implicit methods involve more complex algorithms but are often more powerful for treating systems with disparate time scale dynamics, as discussed below.

The development of symplectic integrators has involved significant interplay among mathematicians, physicists, and chemists. Seminal work on symplectic integrators was done by both physicists and mathematicians (Ruth, 1983; Feng, 1986; Candy and Rozmus, 1991; McLachlan and Atela, 1992; Okunbor and Skeel, 1992; Calvo and Sanz-Serna, 1993) based on second- and third-order explicit approaches and Runge-Kutta methods. Implicit approaches were developed in parallel (Channell and Scovel, 1990; De Frutos and Sanz-Serna, 1992). Recently, these ideas have found their way into the chemistry community (Gray et al., 1994) with promising results. The Verlet integrator (Verlet, 1967), already in common use, was found to be symplectic, thereby explaining the good associated stability observed in practice. However, symplectic integrators that improve on previously available methods have also been developed (Gray et al., 1994). Initial applications using these methods suggest that they may become favored for simulations of polymer dynamics and related problems with small time steps.

The Time Step Problem in Molecular Dynamics

Although standard explicit schemes, such as the Verlet and related methods, are simple to formulate and fast to propagate, they impose a severe constraint on the maximum time step possible. Instability—uncontrollable growth of coordinates and velocities—occurs for step sizes much larger than 1 femtosecond (10^{-15} second). This step size is determined by the period associated with high-frequency modes present in all macromolecules, and it contrasts with the much longer time scales (up to 10^2 seconds) that govern key conformational changes (e.g., folding) in macromolecules. This disparity in time scales urges the development of methods that increase the time step for biomolecular simulations. Even if the stability of the numerical formulation can be ensured, an important issue concerning the reliability of the results arises, since vibrational modes in molecular systems are intricately coupled.

Standard techniques of effectively freezing the fast vibrational modes by a constrained formulation (Ryckaert et al., 1977; van Gunsteren and Berendsen, 1977; van Gunsteren, 1980; Miyamoto and Kollman, 1992) increase the time step by a small factor such as two, still with added complexity at each step. The multiple time step approaches for updating the slow and fast forces provide additional speedup (Streett et al., 1978; Grubmüller et al., 1991; Tuckerman and Berne, 1992; Watanabe and Karplus, 1993), although some stability issues are also involved (Biesiadecki and Skeel, 1993).

Implicit Integration Schemes

There are well-known numerical techniques for solving differential equations describing physical processes with multiple time scales (Gear, 1971; Dahlquist and Björck, 1974). Various implicit formulations are available that balance stability, accuracy, and complexity. However, the standard implicit techniques used by numerical analysts (Kreiss, 1991) have not been directly applicable to MD simulations of macromolecules, for the following reasons.

First, such implicit schemes are often designed to suppress the rapidly decaying component of the motion. This is a valid approach when the contribution of these components becomes negligible for sufficiently long times, as is the case for the second term in $y(t) = \exp(-t) + \exp(-100t)$. However, this situation does not hold for biomolecular systems because of the intricate vibrational coupling. It is well recognized that concerted conformational transitions (e.g., in hinge-bending proteins) require a *cooperative* mechanism driven by small-scale fluctuations to make possible a large-scale collective displacement. Thus, although the absence of the positional fluctuations associated with these high-frequency modes may not by itself be a severe problem, the absence of the *energies* associated with these modes may be undesirable for proteins and nucleic acids, since cooperative motions among the correlated vibrational modes may rely on energy transfer from these high-frequency modes.

Second, implicit schemes with known high stability (e.g., implicit Euler) can introduce numerical damping (Zhang and Schlick, 1993). This has prompted the application of such implicit schemes to the Langevin dynamics formulation, which involves frictional and Gaussian random forces in addition to the systematic force to mimic molecular collisions and therefore a thermal reservoir. This stabilizes implicit discretizations and can be used to quench high-frequency vibrational modes (Peskin and Schlick, 1989; Schlick and Peskin, 1989), but unphysical increased rigidity can result (Zhang and Schlick, 1993). Therefore, more rigorous approaches are required to resolve the subdynamics correctly, such as by combining normal-mode techniques with implicit integration (Zhang and Schlick, 1994); significant linear algebra work in the spectral decomposition is necessary for feasibility for macromolecular systems. For example, banded structures for the Hessian approximation (see related discussion in the section on multivariate minimization beginning on page 68) can be exploited in the linearized equations of motion. There has also been some work on implicit schemes that do not have inherent damping, but preliminary experience suggests that for nonlinear systems, desirable energy conservation properties can be obtained only up to moderate time steps (Simo et al., 1992; Zhang and Schlick, 1995). In particular, serious resonance problems have been noted (Mandziuk and Schlick, 1995).

Third, implicit schemes for multiple time scale problems increase complexity, since they involve solution of a nonlinear system or minimization of a nonlinear function at each time step. Therefore, very efficient implementations of these additional computations are necessary, and even then, computational gain (with respect to standard "brute-force" integrations at small time steps) can be realized only at very large time steps.

Future Prospects

The preceding subsections have described several recent accomplishments in the development of integration methods in MD simulations and have also outlined several important challenges for the future. What makes these integration problems particularly challenging is the fact that solutions demand much more than straightforward application of standard mathematical techniques. At this point it appears that the optimal algorithms for MD will require a combination of methods and strategies discussed above, including symplectic and implicit numerical integration schemes that have minimal intrinsic damping, and correct resolution of the subdynamics of the system by some other technique (e.g., normal-mode analysis). Undoubtedly, high-performance implementations will make possible a gain of several orders of magnitude in the simulation times, and there are certainly additional gains to be achieved by clever programming strategies.

References

Allen, M.P., and D.J. Tildesley, 1990, *Computer Simulation of Liquids*, Oxford University Press, New York.

Biesiadecki, J.J., and R.D. Skeel, 1993, Dangers of multiple-time-step methods, *J. Comput. Phys.* 109:318-328.

Calvo, M.P., and J.M. Sanz-Serna, 1993, The development of variable-step symplectic integrators, with application to the two-body problem, *SIAM J. Sci. Comput.* 14:936.

Candy, J., and W. Rozmus, 1991, A symplectic integration algorithm for separable Hamiltonian functions, *J. Comput. Phys.* 92:230.

Channell, P.J., and J.C. Scovel, 1990, A symplectic integration of Hamiltonian systems, *Nonlinearity* 3:231.

Dahlquist, G., and Å. Björck, 1974, *Numerical Methods,* Prentice Hall, Englewood Cliffs, N.J.

De Frutos, J., and J.M. Sanz-Serna, 1992, An easily implementable fourth-order method for the time integration of wave problems, *J. Comput. Phys.* 103:160.

Feng, K., 1986, Difference schemes for Hamiltonian formalism and symplectic geometry, *J. Comput. Math.* 4:279–289.

Gear, C.W., 1971, *Numerical Initial Value Problems in Ordinary Differential Equations*, Prentice Hall, Englewood Cliffs, N.J.

Gray, S.K., and J.M. Verosky, 1994, Classical Hamiltonian structures in wave packet dynamics, *J. Chem. Phys.* 100:5011–5022.

Gray, S.K., D.W. Noid, and B.G. Sumpter, 1994, Symplectic integrators for large scale molecular dynamics simulations: A comparison of several explicit methods, *J. Chem. Phys.* 101:4062–4072.

Grubmüller, H., H. Heller, A. Windemuth, and K. Schulten, 1991, Generalized verlet algorithm for efficient molecular dynamics simulations with long-range interactions, *Mol. Simul.* 6:121–142.

Kreiss, H.-O., 1991, Problems with different time scales, *Acta Numerica* 1:101–139.

Mandziuk, M., and T. Schlick, 1995, Resonance in chemical-system dynamics simulated by the implicit-midpoint scheme, *Chem. Phys. Lett.,* in press.

McCammon, J.A., and S.C. Harvey, 1987, *Dynamics of Proteins and Nucleic Acids*, Cambridge University

Press, Cambridge.

McLachlan R.I., and P. Atela, 1992, The accuracy of symplectic integrators, *Nonlinearity* 5:541.

Meiss, J.D., 1992, Symplectic maps, variational principles, and transport, *Rev. Mod. Phys.* 64:795.

Miyamoto, S., and P.A. Kollman, 1992, SETTLE: An analytical version of the SHAKE and RATTLE algorithm for rigid water models, *J. Comput. Chem.* 13:952–962.

Okunbor, D.I., and R.D. Skeel, 1992, Canonical numerical methods for molecular dynamics simulations, *Math. Com.* 59:439.

Peskin, C.S., and T. Schlick, 1989, Molecular dynamics by the backward-Euler method, *Commun. Pure Appl. Math.* 42:1001–1031.

Ruth, R.D., 1983, A canonical integration technique, *IEEE Trans. Nucl. Sci.* NS-30:2669.

Ryckaert, J.P., G. Ciccotti, and H.J.C. Berendsen, 1977, Numerical integration of the cartesian equations of motion of a system with constraints: Molecular dynamics of *n*-alkanes, *J. Comput. Phys.* 23:327–341.

Schlick, T., and C.S. Peskin, 1989, Can classical equations simulate quantum-mechanical behavior? A molecular dynamics investigation of a diatomic molecule with a Morse potential, *Commun. Pure Appl. Math.* 42:1141–1163.

Simo, J.C., N. Tarnow, and K.K Wong, 1992, Exact energy-momentum conserving algorithms and symplectic schemes for nonlinear dynamics, *Computer Methods in Applied Mechanics and Engineering* 100:63–116.

Streett, W.B., D.J. Tildesley, and G. Saville, 1978, Multiple time step methods in molecular dynamics, *Mol. Phys.* 35:639–648.

Stuart, A.M., and A.R. Humphries, 1994, Model problems in numerical stability theory for initial value problems, *SIAM Review* 36:226–257.

Tuckerman, M.E., and B.J. Berne, 1992, Molecular dynamics in systems with multiple time scales: systems with stiff and soft degrees of freedom and with short and long range forces, *J. Comput. Chem.* 95:8362–8364.

van Gunsteren, W.F., 1980, Constrained dynamics of flexible molecules, *Mol. Phys.* 40:1015-1019.

van Gunsteren, W.F., and H.J.C. Berendsen, 1977, Algorithms for macromolecular dynamics and constraint dynamics, *Mol. Phys.* 34:1311–1327.

Verlet, L., 1967, Computer "experiments" on classical fluids. I. Thermodynamical properties of Lennard-Jones molecules, *Phys. Rev.* 159:98.

Watanabe, M., and M. Karplus, 1993, Dynamics of molecules with internal degrees of freedom by multiple time-step methods, *J. Chem. Phys.* 99:8063–8074.

Zhang, G., and T. Schlick, 1993, LIN: A new algorithm to simulate the dynamics of biomolecules by combining implicit-integration and normal mode techniques, *J. Comput. Chem.* 14:1212–1233.

Zhang, G., and T. Schlick, 1994, The Langevin/implicit-Euler/normal-mode scheme (LIN) for molecular dynamics at large time steps, *J. Chem. Phys.* 101:4995–5012.

Zhang, G., and T. Schlick, 1995, Implicit discretization schemes for Langevin dynamics, *Mol. Phys.*, in press.

mente con los dedos en el cristal, pero ella no lo oyó. Sin mucha prisa, el detective salió de la estación de radio.

En su casa el aparato estaba sintonizado en la XEKA.

—...las barreras que permiten que extendiendo un dedo podamos tocarnos y dejar de ser unos y otros... Aunque sólo sea para poder contarnos una historia. Como la historia que quiso contarnos Virginia hace una semana y que no pudo contar. ¿Se acuerdan de Virginia, aquella adolescente que asesinaron? Todos ustedes lo habrán leído en los periódicos, ha estado en primera plana por la noticia de la captura del asesino... Virginia que hoy, gracias a la magia de las cintas, está aquí. Detengámosla en el aire pensando en ella, escuchemos su historia. Cuidémonos de una ciudad que amenaza con tragarnos. El silencio es la peor forma de muerte. Te escuchamos, Virginia.

Héctor apagó la radio y luego pateó el aparato, sin furia, con conciencia cívica, como cumpliendo una obligación que había que cumplir. Por más que lo intentaran, la voz de Virginia sonaría vacía. Por mucho que las palabras de Laura trataran de ayudarla, de revivirla, la voz de Virginia sonaría como lo que era: una adolescente muerta.

Una semana después, volvió a repetir el gesto, caminó hacia la radio y la apagó a mitad de una polonesa de Chopin. Le arrimó un suave patín al equipo estereofónico. Fue hacia la cocina buscando un refresco. Estaba cansado, aún le dolían las costillas; por eso, necesitaba cosas seguras: un refresco frío. Cosas seguras: las fotos de la muchacha de la cola de caballo, que estaban ahí, inmóviles, reteniendo un gesto para siempre. La calle que no se había movido, que seguía esperando tras la ventana. Una semana antes, cuando abandonaron el hotel, el "Fantasma" comenzó a llorar. El detective lloró un poco también. No le gustaba el recuerdo de dos tipos llorando tomados del brazo por Tacubaya, uno de ellos con un pañuelo sangriento cubriéndole la nariz, el otro, cargando, como si no pesara nada, una vieja maleta negra. Era un recuerdo extraño, sobre todo porque los vislumbraba en el cine de la memoria, desde lejos, desde afuera.

Se quedó un rato observando las fotos de la muchacha de la cola de caballo: ella bailando twist a los quince años; ella paseando por las islas de CU durante la huelga del 68; ella dándole un vaso de leche a su sobrino. Eran sólo fotos, se dijo. No se engañó en lo más mínimo. No había fotos, había recuerdos, había fantasmas.

Cuando acabó el refresco dejó cuidadosamente el casco en el suelo y fue por un segundo refresco. Siempre somos otros, se dijo. La angustia empezaba a ceder. Se quedó mirando el atardecer. Un sol rojo en una ciudad gris.

Los verdaderos fantasmas, el de una adolescente a la que le habían hecho trampa, y le habían falsificado no sólo un suicidio, sino una despedida. Los fantasmas de a deveras: el del Ángel I, un luchador que caía sobre la lona siempre bien y que le había prometido enseñarle, y el de una mujer llamada Celia, de la que el tipo estuvo enamorado un día, y ambos eternamente perseguidos por el fantasma de Zamudio, vagaban insomnes sin poderse encontrar. Eran historias de amor a medio camino. Inexistentes historias de amor. Puras y pinches, culeras historias de amor derrotadas porque nunca fueron. "Como las mías", informó el su reacio subconsciente.

Se quedó pensando en que, de nuev bíamos perdido otra batalla.

Ciudad de México, primave

Índice

Paco Ignacio Taibo II nació en México en 1949. Sus novelas policiales lo han convertido en un escritor de gran popularidad en su país y fuera de él. *Amorosos fantasmas* ya fue publicada en Alemania, y pronto lo será también en Francia, Italia y Estados Unidos. Además ha dado origen a una película y a un radioteatro.

OTROS TÍTULOS DE ESTA COLECCIÓN:

La sombra del dinosaurio
Pablo De Santis
Historietas de Fabián Slongo
Julián tiene una idea fija: encontrar los fósiles de un dinosaurio.
Y unas cartas antiguas lo empujan al sur y al misterio.

Las botas de Anselmo Soria
Pedro Orgambide
Historietas de Oscar Estévez
La aventura de un chico criado en los fortines, y los prodigios de un largo viaje a Buenos Aires.

Un crimen secundario
Marcelo Birmajer
Historietas de Rafael Segura
Dos estudiantes, Aslamim y Tognini, investigan el extraño robo del Banco Restive.

Pesadilla para hackers
Pablo De Santis
Historietas de Pez
Cinco jóvenes especialistas en computación enloquecen, y un asesinato complica aún más las cosas.

Los miasmas del Plata
Álvaro Gutiérrez Silva
Historietas de Patricia Breccia
Miles de peces muertos flotan en el río. Y no es más que el principio de los planes del Caníbal.

Las llaves del tiempo
Pedro Cazes Camarero
Historietas de Pez
Un amor que parece imposible, y un experimento que puede cambiar a la humanidad.

Derrotado por un muerto
Marcelo Birmajer
Historietas de Patricia Breccia
Aslamin y Tognini vuelven, y con una cabeza cortada que predice el futuro.

Las líneas de la mano
Cristina Siscar
Historietas de Gabriela Forcadell
Una historia de amor escondida en el plano de la ciudad.

Astronauta solo
Pablo De Santis
Historietas de Max Cachimba
Un dibujante de historietas tiene que cruzar la ciudad en la última noche del siglo.

Celebración
Pedro Orgambide
Historietas de Oscar Estévez
El día que la Patria cumple cien años, e1 General y el Chino, un fugitivo, se enfrentan.

El sistema de huida de la cucaracha
Gonzalo Carranza
Historietas de Pez
Un joven científico a quien sólo le interesan sus insectos, se ve envuelto en una trama que involucra a nazis, conspiradores y hasta a un hombre lobo.

El fantasma del Teatro Municipal
Enrique Butti
Historietas de Cuk
En la sala vacía hay una adolescente muerta. Pero no es la primera, y quizás tampoco la última.

Vodka con limón
Aldo Tulián
Historietas de Gabriela Forcadell
Una bella astróloga aparece muerta de un disparo. Y quizás no haya otras pistas que las del complejo idioma de los astros.

Impreso en
A.B.R.N. Producciones Gráficas,
Wenceslao Villafañe 468,
Buenos Aires, Argentina,
en abril de 1994.

CPSIA information can be obtained at www.ICGtesting.com
Printed in the USA
LVOW11s1200190914

404892LV00001B/51/P

N- and V-Representability Problems in Classical Statistical Mechanics

Classical equilibrium statistical mechanics presents a class of unsolved N-representability problems analogous to those in the quantum mechanical regime discussed earlier in this chapter. In this case, N refers to the number of particles (atoms or molecules) present, rather than the number of electrons. The most straightforward version of this classical problem concerns a single-species monatomic system (i.e., spherically symmetric identical particles) and involves the pair correlation function $g(r)$. This nonnegative function of interparticle distance r is defined by the occurrence probability of particle pairs at r, relative to random expectation. Consequently, deviations of $g(r)$ greater than 1 indicate that interparticle interactions have biased the distance distribution to a greater-than-random expectation, while deviations less than 1 indicate the opposite.

For many cases of interest, the interparticle potential energy function V can be regarded as a sum of terms arising from each pair of particles present:

$$V = \sum_{i=1}^{N-1} \sum_{j=i+1}^{N} v(r_{ij}) .$$

The pair potentials $v(r)$ typically are taken to satisfy the following criteria:

(a) $v(r) \to +\infty$ as $r \to 0$;

(b) $v(r)$ is bounded, and is piecewise continuous and differentiable for $r > 0$;

(c) $|v(r)| < C/r^n$ $(C > 0, n > 3)$, for $r > \mathrm{R} > 0$.

Under these circumstances, $g(r)$ plays a special role in the thermodynamic properties of the N-particle system (Hansen and McDonald, 1976). This fundamental quantity appears in closed-form expressions giving the pressure and mean energy at the prevailing temperature and density. Furthermore, it appears in expressions for the X-ray and neutron diffraction patterns for the substance; consequently, these diffraction measurements constitute an experimental means for measuring $g(r)$ for real substances. It should be added that $g(r)$ is also one of the traditional results reported from computer simulations of N-body systems (Ciccotti et al., 1987).

The experimentally, or computationally, adjustable parameters are temperature; particle number density; container size, shape, and boundary conditions; and number N of particles. For most cases of interest, one focuses on the infinite-system limit, where the container size and N diverge, while temperature, number density, and container shape are held constant. The central problem then concerns the mapping between the pair of functions $v(r)$ and $g(r)$, where the latter is interpreted as the infinite-system limit function.

Historically, the fundamental theory of classical systems (particularly in the liquid state) concentrated heavily on prediction of $g(r)$ for a given $v(r)$, that is, the mapping from v to g. This has generated several well-known approximate integral equation predictive theories for $g(r)$, including those conventionally identified in the theoretical chemistry literature by the names Kirkwood (1935), Bogoliubov-Born-Green-Yvon (Born and Green, 1949), Percus and Yevick (1958), and hypernetted chain (van Leeuwen et al., 1959) integral equations, each of which has spawned successor refinements. However, in all cases the respective approximations invoked have, strictly speaking, been uncontrolled. Consequently, the local structure and thermodynamic property predictions based on these various integral equations have had only modest success in describing the dense liquid state,

BOX 4.2 Tutorial on Statistical Mechanics and the Importance of Minima and Saddlepoints in Condensed Matter Systems

A large part of computational chemistry is concerned with the properties of systems at or near thermal equilibrium. The statistics of configurations at thermal equilibrium therefore dominate many of the questions studied by chemists. The principles of the statistical mechanics of equilibrium systems are quite simple to state but are profound and sometimes surprising in their results.

A fundamental postulate of equilibrium statistical mechanics is that in the long run, all states of an isolated system that are consistent with conservation of energy will be observed with equal probability. The thermodynamic quantity S, the entropy, is simply a measure of the number Ω of these equally probable states that the system might access, $S = k_B \log \Omega$. Here k_B is Boltzmann's constant. Thus, combinatorial and various counting problems play a special role in our thinking about the thermodynamics of chemical systems. Although each of the states of an isolated system is equally probable, this is not the case when we consider only a part of a system. When only part of a system is being examined, we must ask the question, How many states of the entire system are accessible when a subsystem is in a given configuration? The answer is given by

$$\Omega(\boldsymbol{X}) = e^{S(\boldsymbol{X})/k_B},$$

where $\boldsymbol{X}$ refers to the specified subsystem and the entropy refers to the subsystem's environment.

A very interesting and powerful special case of this formula is used constantly in equilibrium statistical mechanics. If the subsystem considered is only weakly coupled to the rest of a much larger system, we can decompose the total energy of the entire system into parts:

$$E_{TOT} = E(\boldsymbol{X}) + E_{environment}\,.$$

The energetic coupling is small and can be neglected if we are considering a system that is itself fairly large and therefore has a relatively small surface of interaction with the remainder of the system. In this case the counting problem can be solved since we know that the entropy of the environment is a smooth function of its total energy. This then gives a count of states expressed by

$$\Omega(\boldsymbol{X}) = \exp\left[\frac{1}{k_B} S(E_{TOT} - E(\boldsymbol{X}))\right] = \exp\left[\frac{1}{k_B}(S E_{TOT} - E(\boldsymbol{X})\frac{\partial S}{\partial E}\Big|_{E_{TOT}}\right].$$

The probability then of an exactly specified state of a subsystem that is part of a larger one is proportional to this number of states. It is given by the Boltzmann distribution law

$$P(\boldsymbol{X}) = \frac{1}{Z} e^{-E(\boldsymbol{X})/k_B T}.$$

The temperature entering here is the thermodynamic derivative of the entropy and is proportional to the average kinetic energy of each particle in the system. This distribution law contains within it many of the great phenomena of chemistry and physics. First we see that the most important states are those that have the lowest energy. If the energy then is a continuous function of the coordinates of part of the system, the most probable configurations are those that give the *minima* of this potential. Indeed, the coefficient $1/k_B T$ in the Boltzmann distribution law ensures that at the lowest temperatures only the deepest or global minima are important. Chemistry is usually a low-temperature phenomenon–most chemical reactions are studied around room temperature, although, of course, many do occur under greater extremes of conditions–and room temperature corresponds to only one-fortieth of the typical energy scale of chemical bonds. Thus, the Boltzmann distribution law tells us that chemistry will mostly concern itself with the specific configurations that minimize the energy.

Of course, if molecular systems remained entirely at their energy minima, little would go on. Occasionally, a molecular system must make a transition between one minimum on the energy surface and another. To do this, the system must occasionally find itself in an intermediate high-energy configuration, which the Boltzmann distribution law tells us is rather improbable. If we ask which of the relatively improbable intermediate states between two minima are the most probable, it is clear that these should correspond to saddlepoints of the energy. These saddlepoint configurations are known as transition states to chemists. The probability of a system being found at a transition state determines the rate of a chemical transformation. We see, therefore, that the geometry of minima and saddlepoints of potential energy surfaces is extremely important in determining the chemical properties of a molecular system.

Sometimes only certain aspects of a system are considered explicitly. For example, when we study the shapes, structures, and motions of a biological molecule (e.g., a protein immersed in water), we are interested only secondarily in the configurations of the water molecules around this macromolecule.

A special case of these geometrical problems arises when the subsystem being considered is itself rather large and involves strong interactions between its molecular subunits. In this case, it sometimes happens that the minimum-energy saddlepoint actually possesses an extremely high energy. We then say that the transformation between two minima has a large barrier and the transformation will be extremely slow. Sometimes as the subsystem studied grows larger and larger, the transformation barrier itself also grows larger and larger. Thus, for a macroscopic system, certain transformations may actually take place effectively only on infinite time scales. We can then treat each part of the configuration space very nearly as separate regions. This situation arises when a phase transition occurs. The theory of phase transitions is then concerned with the problem of how a many-dimensional configurational space gets fragmented into parts that are separated by very high energy barriers.

The Boltzmann distribution law applies only to a completely specified subsystem that is interacting weakly with its environment. The biological macromolecule is interacting strongly with its solvent environment, and so the Boltzmann distribution law using the energy alone is inappropriate for describing its configurations. On the other hand, for different configurations of the biomolecule, we can in principle compute the number of configurations of the surrounding solvent that are compatible with that configuration of the biomolecule. Thus, the probability of a particular configuration of the biomolecule would have the form

$$P(X) = \exp[S_{env}(X)/k_B]\exp[-E(X)/(k_BT)] = \exp[-F(X)/(k_BT)],$$

where the probability has been rewritten in a Boltzmann-like form in which the energy of the molecular system is combined with the entropy in its environment to form a free energy $F(\mathbf{x}) = E(\mathbf{X}) - TS(\mathbf{X})$, which gives the probability of the subsystem's configuration. For this reason, the geometry of free energy surfaces is often also of great interest to chemists and physicists.

Occasionally the distinction between energies and free energies is blurred in offhand writing by chemists and physicists, and the uninitiated reader must be careful about these distinctions when applications are made.

they have failed to predict the so-called nonclassical singular behavior at the liquid-vapor critical point (Widom, 1965), and they have been largely useless for the study of freezing and melting transitions. Perhaps as a result of these shortcomings, the recent trend in classical statistical mechanics has been to rely heavily on direct computer simulation of condensed-phase phenomena. Because these simulations often require massive computational resources, a case can be made that revival of analytic predictive theory for $g(r)$ would be favorable from the point of view of the "productivity issue" in theoretical and computational chemistry.

In some respects, the inverse mapping of g to v is even more subtle, intriguing, and mathematically challenging. At the outset, one encounters the obvious matter of defining the space of functions $g(r)$ that in fact can be generated by a pairwise additive potential energy function V. A few necessary conditions are straightforward; as already remarked, $g(r)$ cannot be negative. It is generally accepted (but not rigorously demonstrated) that g must approach unity as r diverges if the temperature is positive, even though the system itself may be in a spatially periodic crystalline state. In addition, the Fourier transform of $g(r) - 1$,

$$G(k) = \int \exp\,(i\boldsymbol{k}\cdot\boldsymbol{r})\;[g(r) - 1]\;d\boldsymbol{r},$$

is also subject to necessary conditions stemming from the nature of the linear equilibrium response of the system to weak external perturbations: for all $k > 0$ one must have (Percus, 1964)

$$1 + \rho\,G(k) \geq 0 \qquad (\rho = \text{number density}).$$

These generic conditions can be supplemented by others that are necessary if $v(r)$ has an infinitely repelling hard core, that is,

$$\begin{aligned} v(r) &= +\infty \text{ for } 0 < r < a, \\ v(r) &= \text{bounded for } a < r. \end{aligned}$$

This hard-core property prevents neighbors from clustering too densely around any given particle, and from the geometry of hard-sphere close packings it is possible to bound the integral of $r^2 g(r)$ over finite intervals of r.

A primary challenge concerns formulation of sufficient conditions on $g(r)$, given that V possesses the pairwise-additive form displayed above. At present we have no rational criterion for deciding whether a given $g(r)$, however "reasonable" it may appear to be by conventional physical standards, corresponds to the thermal-equilibrium short-range order for *any* pairwise additive V. It is not even clear at present how to construct a counterexample, namely, a $g(r)$ meeting the necessary conditions above that *cannot* map to a $v(r)$ of the class described. In any case, formulation of sufficient conditions would likely improve prospects for more satisfactory integral equation (or other analytical) predictive techniques for $g(r)$.

Several directions of generalization exist for this classical V-representability problem; these include the following matters:

1. Properties of triplet and higher-order correlation functions g^n for occurrence probabilities of particle n-tuples;
2. Properties of correlation functions for particles (molecules) with internal degrees of freedom (rotation, vibration, conformational flexibility);
3. Effects of specific nonadditive potentials, which would be the case when including

three-particle contributions in V; and,

4. Multicomponent (several species, or mixture) systems, in particular the important case of electrostatically charged particles (ions) with their long-ranged Coulombic interactions.

References

Born, M., and H.S. Green, 1949, *A General Kinetic Theory of Liquids*, Cambridge, London.

Ciccotti, G., D. Frankel, and I.R. Kirkwood, 1987, eds., *Simulation of Liquids and Solids*, North-Holland, Amsterdam.

Hansen, J.P., and I.R. McDonald, 1976, *Theory of Simple Liquids*, Section 2.6, Academic Press, New York.

Kirkwood, J.G., 1935, Statistical mechanics of fluid mixtures, *J. Chem. Phys.* 3:300–313.

Percus, J.K., 1964, in *The Equilibrium Theory of Classical Fluids*, H.L. Frisch and J.L. Lebowitz, eds., W.A. Benjamin, New York, pp. II–33 to II–170 (see particularly II–41).

Percus, J.K., and G.J. Yevick, 1958, Analysis of classical statistical mechanics by means of collective coordinates, *Phys. Rev.* 110:1–13.

van Leeuwen, J.M.J., J. Groeneveld, and J. De Boer, 1959, New method for the calculation of the pair correlation function, *Physica* 25:792–808.

Widom, B., 1965, Equation of state in the neighborhood of the critical point, *J. Chem. Phys.* 43:3898–3905.

Implications of Topological Phases

The Born-Oppenheimer approximation dates from the 1920s, and the entire notion of molecular structure can be based upon it. It is thus a surprise that significant qualitative physics has been ignored by most chemical physicists in applying the Born-Oppenheimer approximation to systems with degenerate electronic states. The basic idea behind the Born-Oppenheimer approximation is that nuclei move much more slowly than electrons. Thus, the Schrödinger equation for electrons can be solved at fixed nuclear configuration and the resulting energy can be used as a potential for studying the motions of the nuclei themselves.

Generally, when nuclear motion itself is quantized, one assumes the usual Schrödinger equation with a classical scalar potential for the nuclear motions. This has proved valid for systems that do not have significant electronic degeneracy. A serious mathematical problem is the uniqueness of the wave function for the nuclei. The Born-Oppenheimer approximation really assumes a single path for the slowly moving nuclei. If there is an electronic degeneracy, topologically distinct paths may connect two different positions on the same electronic surface. Thus, in addition to the phases that one develops for the quantum dynamics through the simple scalar potential dynamics, there is an additional topological phase. The existence of this topological phase, which depends on the path between two points, has been known since at least the 1950s, when Longuet-Higgins studied it in the context of Jahn-Teller distortions. Only in recent years has its significance been truly appreciated, however. One of the leaders in bringing out the significance of topology in quantum molecular dynamics was M. Berry. However, it was appreciated somewhat earlier by Truhlar and Mead that this topological phase plays a role in chemical reactions. Indeed it is important even in the most fundamental of chemical reaction problems, the $H + H_2$ reaction. Very recently, the discrepancy

BOX 4.3 Implications of Dynamic Chaos for Quantum Mechanical Systems

Many phenomena in chemistry are at the border of applicability of classical mechanics. Quantum mechanical phenomena, such as tunneling and interference, certainly are relevant to many chemical reactions. Thus, in addition to purely classical dynamical methods, semiclassical approximations are used quite commonly in chemical physics.

Semiclassical methods work fairly well for low-dimensional systems such as those encountered in gas-phase chemical reactions, because the collisions that act as randomizers are infrequent and the chaotic character of the processes may often be neglected. On the other hand, in attempting to extend semiclassical methods to condensed-phase systems, one is immediately faced with the problem of the underlying classical chaos. No completely adequate semiclassical quantization of chaotic systems yet exists.

Most of the effort of theoretical chemists working in this area has been devoted to understanding simple themes that may give some qualitative insight to phenomena that occur in the quantum mechanics of systems that are classically chaotic. Several important themes have been developed, one of the most notable being the connection of quantum chaos to random matrix theory. The notions behind this have their roots in Wigner's use of random matrices to describe nuclear systems, but the application of these ideas to molecules has been equally rewarding. One can examine the evolution of the energy levels of a system under perturbation in order to understand its quantum chaotic nature. It has been shown that a random matrix description could arise from multiple level crossings. This approach has also been shown to be related to some exactly integrable systems of particles moving in one dimension. The great irony is that the random matrix description arises from a problem that, in another guise, leads to exactly integrable equations.

Very interesting connections exist to the theory of solutions of nonlinear partial differential equations. Classical systems in many dimensions, when they are chaotic, often exhibit diffusive dynamics. Arnold has shown how weakly coupled systems of dimension higher than 2 exhibit such diffusion. It has recently been argued that a phenomenon analogous to Arnold diffusion in the classical limit arises in quantum problems and that weakly coupled systems of quantized oscillators are analogous to local random matrix models. These local random matrix models are closely tied to the problem of Anderson localization, which concerns itself with the nature of eigenfunctions of random differential operators.

A most enticing development in the understanding of quantum chaos has been the connection of problems in quantum chaos with deep problems in number theory. One of the most picturesque approaches for obtaining quantum mechanical energy levels is to calculate the Green's function of the Schrödinger equation through a sum over classical paths. For chaotic systems these classical paths are extremely numerous and the Green's function is indeed the sum of a statistically fluctuating quantity that itself presents interesting mathematical problems to be discussed later.

One model for the classical paths represents them as repetitions of some fundamental periodic orbits. Some very simple special models of the actions of these orbits lead to a Green's function that is closely tied to the Riemann zeta function. The prime numbers represent the fundamental periodic orbits. This has suggested that the zeros of the Riemann zeta function are related to the quantum mechanical eigenvalues of some Hamiltonian that is classically chaotic. This ansatz has led to interesting predictions about the spacing of zeros of the Riemann zeta function and other statistics that seem very much in keeping with the random matrix theories being used to describe quantum chaos. Thus, it seems that problems in quantum chaos might be clarified considerably by considerations from probabilistic number theory and, conversely, that deep number theoretic questions might be addressed by using ideas from the quantum mechanics of chaotic systems. Although significant progress has already been made in developing conjectures based on these ideas, there is still a tremendous amount to do and many deep mysteries remain.

between experimental results for H + H_2 and large-scale computations of the scattering cross sections was shown to arise from neglect of this topological phase.

For problems with small amounts of degeneracy, the topological phase is easy to handle with little mathematical sophistication. Either a trajectory encircles a conical intersection (of Born-Oppenheimer energy surfaces) or it does not, leading to two values of the phase. This encircling of singularities can be described by using the idea of a gauge potential. With higher degeneracies, however, very difficult topological problems may be encountered since many surfaces can make avoided crossings in many locations. The paradigm of such complicated topology problems may well be metal clusters. For metals in the thermodynamic limit, there are numerous energy levels corresponding to the excitation of electrons just below the Fermi sea to just above it. Since the electronic levels are highly delocalized, these energy changes are quite small and the energy surfaces are close together. The actual dynamics of the nuclei must involve the coupling of several surfaces. There are many possible interchanges of the metallic ionic cores, and complicated topologies can result.

Another place in which topology enters is when an underlying approximate wave function is built up out of many degenerate electronic wave functions and the dynamics of electronic excitations is studied. The paradigm for this is the recent interest in resonating valence bond descriptions of metallic and superconducting materials. Here, reorganization of the different valence bond structures as an excited electron or hole moves around gives rise to topological phases and gauge fields. It has been argued that these effects are at the heart of the new high-temperature superconductors and represent a real breakdown of the traditional band structure picture of metals. Most models studied by physicists, however, have been very simple, and it will be necessary to understand how the topological phases arise in completely realistic electronic structure calculations if one is to make predictions of new high-temperature superconductors on the basis of these ideas.

Theoretical and Computational Chemistry in Spaces of Noninteger Dimension

A major mathematical landmark in the eighteenth century was Euler's introduction and exploitation of the famous gamma function. One of its basic and striking properties is that it provides a natural "smooth" extension of the factorials *n*! that are defined nominally just for the positive integers to all positive numbers, and indeed even into the complex plane. The pervasive appearance of the Euler gamma function throughout classical mathematical analysis constitutes a powerful paradigm suggesting that analogous extensions from the discrete positive integers to the complex plane in other contexts might generate analogous intellectual benefits.

During roughly the last two decades, simultaneous developments in several distinct areas of physical science appear to point to the necessity (or at least the desirability) of just such an extension. Specifically, this involves generalizing the familiar notion of Euclidean *D*-dimensional spaces from positive integer *D* at least to the positive reals, if not to the complex *D*-plane. This is not an empty pedantic exercise; at least one serious proposal has been published (Zeilinger and Svozil, 1985) claiming that accurate spectroscopic measurements of the electron "g-factor" indicate that the space dimension of our world is less than 3 by approximately 5×10^{-7}. Furthermore, in various theoretical applications that have so far been suggested for the continuous-*D* concept, *D* itself or its inverse appears to be a natural expansion parameter for various fundamental quantities of interest. However, most of the work along these lines thus far has been ad hoc, lacking rigorous mathematical underpinning. Naturally this calls into question the validity of claimed results.

Three physical science research areas deserve mention in this context. The first is quantum field theory; dimension *D* has been treated as a continuously variable "regularizing parameter" whose manipulation avoids embarrassing divergences in perturbation expansions (Bollini and Giambiagi, 1972; t'Hooft and Veltman, 1972; Ashmore, 1973). The second is the statistical mechanics of phase transitions (specifically involving critical point phenomena); because of rigorously known results for $D = 2$ and $D = 4, 5, 6, \ldots$, series expansions in the quantity $4-D$ have been developed for various quantities of interest to access the physical case $D = 3$ (Wilson and Fisher, 1972; Gorishny et al., 1984). The third area holds perhaps the greatest promise for chemical progress, namely, the development of atomic and molecular quantum mechanics (with useful computational algorithms) in spaces of arbitrary D (Goodson et al., 1992; Herschbach et al., 1992).

As in the other applications, the notion of atomic and molecular quantum mechanics is unambiguously defined for D a positive integer; in other words, the Schrödinger wave equation and

BOX 4.4 Nodal Properties of Wavefunctions

Knowledge of the nodes of the many-fermion wavefunction would make possible exact calculation of the properties of fermion systems by Monte Carlo methods. Little is known about nodes of many-body fermion systems even though the one-dimensional case is ubiquitous in textbooks on quantum mechanics. The nodes referred to here are the nodes of the exact many-body wavefunction and are very different from the nodes of orbitals.

In the absence of a rigorous simulation method for fermion systems, the fixed-node approximation has been found to be a useful and powerful approach. One assumes knowledge of where the exact wavefunction is positive and negative based on the nodes of a trial wavefunction. The Schrödinger equation in imaginary time is solved by simulating the diffusion process with branching within the regions bounded by the assumed nodes.

For the ground state, Ceperley (1991) has proved that ground state nodal cells have the tiling property (i.e., there is only one type of nodal cell, all other cells being related by permutational symmetry). The tiling property is the generalization to fermions of the theorem that a bosonic ground state is nodeless.

The nodal hypervolumes of a series of atomic N-body Hartree-Fock level electronic wavefunctions have been mapped by using a Monte Carlo simulation in $3N$-dimensional configuration space (Glauser et al., 1992). The basic structural elements of the domain of atomic and molecular wavefunctions have been identified as nodal regions and permutational cells (identical building blocks). The results of this study on lithium-carbon indicate that Hartree-Fock wavefunctions generally consist of four equivalent nodal regions (two positive and two negative), each constructed from one or more permutational cells.

A generalization of the fixed-node method has been proposed that could solve the fermion problem at finite temperature if only the nodes of the fermion density matrix were known (Ceperley, 1991).

References

Ceperley, D.M., 1991, *J. Stat. Phys.* 63:1237.

Glauser, W.A., W.R. Brown, W.A. Lester, Jr., D. Bressanini, B.L. Hammond, and M.L. Koszykowski, 1992, *J. Chem. Phys.* 97:9200.

its boundary conditions have an immediate and clear meaning. The desire to embed these problems in the arbitrary-D context arises primarily from the observation that solutions to the Schrödinger equation adopt a simple limiting form as D approaches infinity, namely, those for simple harmonic oscillators localized in multidimensional space (Goodson et al., 1992; Herschbach et al., 1992). Eigenfunction and eigenvalue expansions in $1/D$ have then been formally generated, with the hope that series summation techniques (e.g., Padé approximants) would permit extension to the case of ultimate interest $D = 3$. This strategic approach to real chemistry in the real world is emboldened by the facts that (a) $D = 1$ is often an exactly solvable case (or at least amenable to very accurate numerical study), and (b) exact interdimensional identities for D and $D + 2$ are known (Herrick, 1975). These latter afford convenient fixed points for refining the series summation attempts.

The presumption that spaces with noninteger dimension were available as analytic tools for atomic and molecular quantum mechanics rests largely on simple observations such as the fact that the D-dimensional (hyper)spherical volume element,

$$dV(D)/dr = K(D)r^{(D-1)}$$

$$K(D) = 2\pi^{(D/2)}/\Gamma(D/2)\,,$$

is an obvious analytic function of the variable D. The implicit assumption in the various applications to date, quantum mechanical and otherwise, seems to have been that the same expression can be invested with mathematical legitimacy for noninteger D, in the sense that it is an attribute of a family of precisely defined spaces. This is by no means an obvious proposition, since any quantity such as $K(D)$ above could be augmented by any function of D that vanishes at the positive integers, such as $\sin(2\pi D)$, without affecting the situation for conventional Euclidean geometry.

The published literature reveals some attempts to axiomatize spaces of noninteger dimension (Wilson, 1973; Stillinger, 1977), but it is clear that the subject requires deeper mathematical insight than it has thus far experienced. In particular, it is desirable to determine the extent to which arbitrary-D spaces are uniquely definable as uniform and isotropic metric spaces and what their relation to conventional vector spaces might be. It has been suggested (Wilson, 1973) that noninteger-D spaces can be viewed as embedded in an infinite-dimensional vector space, but whether this is uniquely possible or even necessary to perform calculations remains open.

It is important to stress the distinction between the general-D spaces that may be obtained by interpolation between the familiar Euclidian spaces for integer D on the one hand and the so-called fractal sets to which a generally noninteger Hausdorff-Besicovitch dimension can be assigned (Mandelbrot, 1983). The latter are normally viewed as point sets contained in a Euclidean host space; furthermore, they fail to display translational and rotational invariance, and are therefore not uniform and isotropic.

References

Ashmore, J.F., 1973, On renormalization and complex space-time dimensions, *Commun. Math. Phys.* 29:177–187.

Bollini, C.G., and J.J. Giambiagi, 1972, Dimensional renormalization: The number of dimensions as a regularizing parameter, *Nuovo Cimento B* 12:20.

Goodson, D.Z., M. Lopez-Cabrera, D.R. Herschbach, and J.D. Morgan III, 1992, Large-order dimensional perturbation theory for two-electron atoms, *J. Chem. Phys.* 97:8481.

Gorishny, S.G., S.A. Larin, and F.V. Tkachov, 1984, *Phys. Lett.* 101A:120.

Herrick, D.R., 1975, Degeneracies in energy levels of quantum systems of variable dimensionality, *J. Math. Phys.* 16:281.

Herschbach, D.R., J. Avery, and O. Goscinski, eds., 1992, *Dimensional Scaling in Chemical Physics*, Kluwer Academic, Dordrecht, Holland.

Mandelbrot, B.B., 1983, *The Fractal Geometry of Nature*, W.H. Freeman, San Francisco.

Stillinger, F.H., 1977, Axiomatic basis for spaces with noninteger, *J. Math. Phys.* 18:1224.

t'Hooft, G., and M. Veltman, 1972, *Nuclear Phys. B* 44:189.

Wilson, K.G., 1973, *Phys. Rev. D* 7:2911.

Wilson, K.G., and M.E. Fisher, 1972, *Phys. Rev. Lett.* 28:240.

Zeilinger, A., and K. Svozil, Measuring the dimension of spacetime, *Phys. Rev. Lett.* 54:2553.

Multivariate Minimization in Computational Chemistry

Introduction

Mathematical optimization is a branch of mathematics that seeks to answer the question "What is best?" for problems in which the quality of the answer can be expressed as a numerical value (see, e.g., Gill et al., 1983b; Fletcher, 1987; Ciarlet, 1989). This question might refer to the "best" approximation in some local sense (i.e., a local solution) or to the global solution over the entire feasible space (i.e., the global minimum) (see, e.g., Nemhauser et al., 1989; Floudas and Pardalos, 1991). A common problem arises when a complex physical system is described by a collection of particles, or combinations of states, in a multidimensional phase space. An energy or cost function is associated with each different configuration, and the challenge is to find sets of points that minimize (or maximize)[1] the objective function. Such applications arise frequently in molecular modeling, rational drug design, quantum mechanical calculations, mathematical biology models, neural networks, combinatorial problems, financial investment planning, engineering, electronics, meteorology, and computational geometry. In applications that arise in computational chemistry (Scheraga, 1992; Schlick, 1992), the feasible space is often very high in dimensionality and complexity, so both local and global minima are of interest.

There are many optimization techniques available for the computational scientist. Nonetheless, implementation of the more sophisticated techniques requires considerable computing experience, algorithm familiarity, and intuition. While software vendors offer a variety of "black-box" codes, serious practitioners frequently discover that a good deal of understanding and modification is required for successful applications. Such modifications involve tailoring the algorithm to features of the problem at hand—such as function separability—or exploiting available experimental information that might guide the optimization path—such as nuclear magnetic resonance (NMR) distance restraints in molecular mechanics. Moreover, successful new optimization schemes may be not be known or available to nonspecialist mathematicians, let alone to scientists in allied fields.

Thus, the transfer of knowledge, its application to real problems, and its further developments will greatly benefit from increased interdisciplinary interactions. In particular, it may be useful to

[1]These are equivalent problems. The minimum of a function f is the maximum of the function $-f$.

stimulate algorithmic developments in optimization toward important scientific problems, such as arise in chemistry, that would involve synergistic efforts on both parts: the application-oriented scientist and the algorithm developer. Such collaborations are likely to be fruitful to both parties, since testing of new methods will be possible on real-life problems and might generate an evolving body of solutions that take into account the available physical data. There is recent evidence (e.g., in special sessions of meetings of the Society for Industrial and Applied Mathematics) that mathematicians have discovered the challenges in "mathematical chemistry problems" and protein folding, but many frontiers lie ahead.

Problem Classification

The available optimization algorithms are classified according to the features of the target problem. The objective function to be minimized (or maximized) may be formulated in terms of *integer* variables (discrete optimization), *integers in permutations* (combinatorial optimization), continuous real numbers (continuous optimization), or *both continuous real numbers and integers* (mixed integer optimization). Examples from these four classes involve, respectively, order planning for organizations (the integers may denote, for example, the number of units of each item to be purchased monthly for a restaurant); the traveling salesman problem (the ordered list of N integers represents a cyclical itinerary for visiting N cities); molecular structure prediction (the real numbers may denote nuclear or electronic positions of the particles, or a set of internal variables describing the molecular system); and airline crew scheduling (the integers may identify particular flight routes and the real numbers may refer to the hours of shift for the flight crew). For computational chemistry, continuous optimization is the most important type of problem.

In addition to the nature of the control (or independent) variables, the objective function may be linear, quadratic, or nonlinear (the latter in varying extent). The problem may be formulated as unconstrained or constrained, with constraints involving equality or inequality conditions, which may be linear, quadratic, or nonlinear. Thus, for the above examples, constrained formulations may introduce upper and lower bounds for the Cartesian positions or specified values for certain internal variables that should remain fixed; the airline crew scheduling problem will incorporate into the optimization formulation the total number of scheduled flights, lower and upper bounds for the lengths of shifts, enforced limits on gaps between transatlantic flights, and so on. In addition to functional form and constraints, other important considerations involve the cost of evaluating the objective function and the availability (or lack) and associated cost of derivatives. In some cases, the derivatives may be discontinuous, and special techniques may be required. Derivative information can often be exploited significantly for the optimization algorithm, but the benefits must be balanced with the additional costs involved.

The Complexity of Computational Chemistry Problems

Optimization problems frequently arise in molecular and quantum mechanical calculations in chemistry. These problems are typical of optimization applications seeking favorable configurational states of a physical system. The large-scale nature of these problems together with the lack of convexity rules out exhaustive sampling in the feasible space except for very small systems. Therefore, clever optimization methods are a necessity, and their improvement translates into the ability to model larger physical systems and generate important structural predictions.

The expense of calculating the function and the associated derivatives also introduces difficulties that limit the type of algorithm that may be utilized. In many molecular mechanics applications, it may be tedious but possible to calculate the derivatives; often, the additional computational cost involved in computing the gradient is only a small factor more (e.g., 4 to 5) than computing the function (and guaranteed by automatic differentiation, which also saves coding efforts; see Box 4.5).

Computational intensity often stems from the long-range interactions among the N particles in the system (e.g., Coulombic forces). In molecular mechanics, the *direct* evaluation requires on the order of N^2 operations, and even if a cutoff radius is introduced, computation of the nonbonded terms dominates computation time. Implementation of fast particle methods (Greengard, 1994) in molecular mechanics and dynamics calculations (Grubmüller et al., 1991; Board et al., 1992) is clearly important for reducing the severity of this problem and allowing more accurate representation of the

BOX 4.5 Automatic Differentiation

Automatic differentiation, essentially a new algebraic construct (Rall, 1981; Griewank, 1988; Griewank and Corliss, 1991), provides a way to compute exact derivatives of a function whose calculation is expressed by a computer program. This technique may be useful for minimization of computational chemistry problems. Finite-difference techniques have been used for this purpose for a long time, but they require judicious choices in the finite-difference interval (Gill et al., 1983a); even with optimal choices, errors are inevitable in regions where the gradient is very small or in those where function curvature changes very rapidly. Finite differences also become very expensive in large dimensions, unless partial separability is exploited.

There have been some "computer algebra" approaches based on symbolic computations (e.g., packages like Macsyma, Mathematica, or Maple) that can actually handle some differentiation tasks when the functional form is specified. Parenthetically, the reduction in mathematical errors and ready availability of graphics due to such symbolic computing tools have enhanced productivity in a number of areas of chemistry. However, while simplifying the programmer's work considerably, symbolic programs are not practical to use in the context of large-scale computer programs that require repeated evaluation and differentiation of a complex function. Moreover, this approach cannot be applied to full Fortran, C, or C++ programs directly, only to simplified or special syntax.

In contrast to finite differences and symbolic algebra, the more recent approach of automatic differentiation is based on compiler techniques. Specifically, compiler transformations are applied to compute rigorously the derivative of a function defined by a program. The basic idea is to use the chain rule from calculus to compute the derivative of a composition of functions. Mathematically, this is similar to using symbolic algebra, but the chain rule is applied to the numerical intermediate results instead of to the expressions themselves, which makes it much more efficient. Automatic differentiation can be applied to complete programs including common blocks, equivalence statements, GOTO statements, and other features beyond the scope of existing symbolic algebra systems. Any code can be viewed as such a composition of elementary functions and algebraic operations, with the dependencies of one variable upon another being traced with modern compiler techniques. A code may have conditional branches that depend on the values of the independent variable (either explicitly or implicitly). The differentiation software must generate corresponding conditional branches. Originally, packages were developed for programs written in C and C++, but recent efforts have extended this to include Fortran programs as well (Bischof et al., 1992).

Because the derivatives are expressed in exact symbolic form, their calculation is subject only to rounding errors. However, if a function is nondifferentiable at some point, it is not entirely clear what the result of automatic differentiation will be; in many cases, it corresponds to the derivative computed from one side of the nondifferentiability. If a function is defined by table-lookup, then the derivative returned by automatic differentiation may well be zero. Some cases of anomalous behavior in differentiating specific codes are the subject of ongoing research.

Significantly, automatic differentiation techniques turn out to be competitive with the finite difference approach computationally. Their application to computational chemistry codes is just beginning.

long-range interactions; the advantage of such an approach has already been demonstrated in other scientific applications (Greengard, 1994), for example in the context of integral equations in engineering problems (Nabors et al., 1994).

The multiple-minimum problem is a severe hurdle in many large-scale optimization applications. The state of the art today is such that only for small and reasonable problems do suitable algorithms exist for finding all local minima for linear and nonlinear functions. For larger problems, however, many trials are generally required to find local minima, and finding the global minimum cannot be ensured. These features have prompted research in conformational-search techniques independent of, or in combination with, minimization (Leach, 1991). To illustrate, consider a simple model for an alkane chain of m units (residues). From combinations or rough partitions in favorable structures of the individual building blocks, the number of possible starting points produces 3^m starting configurations. For polypeptides and polynucleotides, the flexibility of the monomer (building block) configurations increases, producing a rough range of 10^m to 25^m reasonable starting points by coarse subdomain partition (e.g., combinations of typical side chain, main chain, backbone, or sugar dihedral angles). Exhaustive searches are clearly not feasible.

> A molecular **configuration** is described by a list of numbers that specifies the *relative* position of the atoms in space. By definition, the configuration is unchanged when the molecule as a whole is subjected to rigid-body motion (translation or rotation). If the molecule consists of N atoms, $3N - 6$ numbers are required to specify its configuration uniquely. These numbers may consist of $3N - 6$ Cartesian positions (with 6 values fixed for uniqueness) or some combination of bond lengths, bond angles, and dihedral angles (angles defining the rotation between two groups with respect to the bond connecting them). The term **conformation** is typically used by chemists to describe the spatial configuration of a molecular system—strictly speaking, one with fixed bond lengths and valence angles.

The *buildup* technique is a related configurational search strategy, used in studies of proteins (Pincus et al., 1982) and nucleic acids (Hingerty et al., 1989). Reasonable starting points are constructed by combining *minima* of conformational building blocks. This rational strategy has performed rather well in practice, but there is no guarantee that all biologically important local minima, much less the global minimum, are revealed. One of the problems is the nonlocal nature of the interactions in the folded macromolecule. That is, segments far apart in the linear sequence will make close contact upon folding; thus, the collective minimum may not correspond to any minima of the constituent building blocks. Furthermore, the number of starting points is still exponential in the number of building blocks. This buildup technique might be an interesting mathematical area to explore further, perhaps through techniques of interval analysis (see below under global optimization methods).

Molecular dynamics, discussed on pages 54–58, can also be viewed as a technique for obtaining structural information (e.g., mean atomic fluctuations, dynamical pathways, isomerization rates) that is complementary to potential energy minimization. While in theory information on all thermally accessible states should be observable, the restriction of the integration time step to a very small value with respect to time scales of collective biomolecular motions limits the scope of molecular dynamics in practice.

Local Optimization Methods

Local methods are defined by an iterative procedure that generates iterates $\{\mathbf{x}_0, \mathbf{x}_1, \ldots, \mathbf{x}_k, \ldots\}$ intended to converge to a local minimum $\mathbf{x}^*$ from a given $\mathbf{x}_0$. Their performance is clearly sensitive to the choice of starting point in addition to search direction and algorithmic details. In the line-search subclass, a search vector $\mathbf{p}_k$ is computed at each step by a given strategy, and the objective function f is minimized approximately along that direction so that "sufficient decrease" is obtained (see, e.g., Dennis and Schnabel, 1983; Luenberger, 1984). In trust-region approaches, a local quadratic model of the function is minimized at every step using current Hessian information, and an optimal step is chosen to lie within the "trust region" of the quadratic model (Dennis and Schnabel, 1983).

Local deterministic optimization methods have experienced extensive development in the last decade (e.g., Nocedal, 1991; Wright, 1991). Studies have produced a range of robust and reliable techniques tailored to problem size, smoothness, complexity, and memory considerations. Many variants of Newton's method have been produced that extend applicability far beyond small or sparse problems. Nonderivative methods are generally not competitive, but significant developments have been made in nonlinear conjugate gradient (CG) methods (generally recommended for very large problems whose function is very expensive to evaluate) and Newton methods.

The classes and extensions of Newton's method, the prototype of second-derivative algorithms, include *discrete* Newton, *quasi*-Newton (QN) (also termed *variable metric*), and *truncated* Newton (TN) (e.g., Dennis and Schnabel, 1983; Gill et al., 1983b). Historically, because of the $O(n^2)$ memory requirements, where n is the number of variables in the objective function, and the $O(n^3)$ computation associated with solving a linear system directly, Newton methods have been most widely used (1) for small problems, (2) for problems with special sparsity patterns, or (3) when near a solution, after a gradient method has been applied. Fortunately, advances in computing technology and algorithmic developments have made the Newton approach feasible for a wide range of problems. Indeed, effective strategies have been tailored to available storage and computation, exhibiting good performance in theory and practice, and this trend will undoubtedly intensify.

Two specific classes are emerging as the most powerful techniques for large-scale applications: *limited-memory* quasi-Newton (LMQN) and *truncated* Newton methods. LMQN methods attempt to retain the modest storage and computational requirements of CG methods while approaching the superlinear convergence properties of standard (i.e., full memory) QN methods (Gilbert and Lemaréchal, 1989; Liu and Nocedal, 1989; Nash and Nocedal, 1991; Zou et al., 1993). Similarly, TN algorithms attempt to retain the rapid quadratic convergence rate of classic Newton methods while making computational requirements feasible for large-scale functions (Dembo and Steihaug, 1983; Nash, 1985; Schlick and Overton, 1987). With advances in automatic differentiation (see Box 4.5), the appeal of these methods will undoubtedly increase even further (Dixon, 1991).

Both limited-memory QN and TN methods are promising for computational chemistry problems. Moreover, they can be adapted to both constrained and unconstrained formulations and can exploit the special composition (distinct components) of the potential energy function to accelerate convergence (Derreumaux et al., 1994). This issue involves a natural separation of the objective function into components of differing complexity (e.g., local and nonlocal interactions). This special composition can be exploited to construct banded or other sparse preconditioners in the context of CG and TN. Such problem tailoring requires some familiarity with the algorithmic modules and also demands knowledge of the theoretical and practical strengths and weaknesses of the different minimization methods. With rapidly growing improvements in high-performance vector and massively parallel machines, application-tailored software may be even more important in combination with parallel architectures whose design is motivated by specific applications.

Global Optimization Methods

In their attempt to find a global rather than local minimum, global optimization methods tend to explore larger regions of function space (see, e.g., Dixon and Szegö, 1975; Floudas and Pardalos, 1991). The global minimum of a function can be sought through two classes of approaches:

deterministic and *stochastic*. Deterministic methods usually require the objective function to satisfy certain smoothness properties; they construct a sequence of points converging to lower and lower local minima. Ideally, they attempt to "tunnel" through local barriers. Local minimization methods are often required repeatedly in the framework; hence, developments in local methods are likely to have an important impact on global techniques as well. Computational effort tends to be very large, and a guarantee of success can be obtained only under specific assumptions.

Stochastic global methods, on the other hand, involve systematic manipulation of randomly selected points (Nemhauser et al., 1989; Rinnooy Kan and Timmer, 1989; Schnabel, 1989; Törn and Žilinskas, 1989; Byrd et al., 1990). Success can be guaranteed only in an asymptotic, stochastic sense, although in practice many applications are very promising.

In the early days of global optimization (mid-1970s), most efforts focused on stochastic or heuristic approaches (Dixon and Szegö, 1975). In chemical applications, simulated annealing (Metropolis et al., 1953; Kirkpatrick et al., 1983; Dekkers and Aarts, 1991) is an appealing method of this class and is effective for small to medium molecular systems. It is also very easy to implement and generally requires no derivative computations. Indeed, there has been a wide application of this method to chemical systems.

More recent efforts have focused also on deterministic global optimization methods. Interesting examples include the tunneling method (Levy and Gomez, 1985; Levy and Montalvo, 1985) and several innovative deterministic approaches in chemical applications (Purisima and Scheraga, 1986; Piela et al., 1989; Scheraga, 1992; Shalloway, 1992). In particular, in the mathematical community, two recent powerful methods have been identified that might be useful to chemical applications. One exploits convex properties and is based on differences of convex functions (Pardalos and Rosen, 1987; Horst and Tuy, 1993); the other is based on interval analysis (Hansen, 1980, 1992; Neumaier, 1990; Schnepper and Stadtherr, 1993). The convexity approach has been successful for global quadratic problems of up to approximately 300 variables and 50 constraints (Pardalos and Rosen, 1987; Horst and Tuy, 1993). Interval analysis, a field little known even to mathematicians, was pioneered by Hansen, among others. It involves computation of strict bounds to bracket the global minimum of a function. The algorithms involve various branch and bound techniques that recursively split the configuration space, aiming at bracketing the minimum as tightly as possible. Other information, such as bounds on derivatives, may also be generated. This class of methods can be applied to the solution of nonlinear systems, as well as global constrained and unconstrained optimization. However, these methods require second-derivative information (Hessians for optimization problems) and, moreover, the inverse of a preconditioning matrix to produce realistic bounds. For these reasons, interval analysis has been applied only to relatively small problems thus far. However, future research may be promising with preconditioning techniques that are now well developed for local optimization.

Perspective

In sum, the optimization applications that arise in computational chemistry offer challenging and rewarding problems to mathematicians. There is a need for the development of both local and global methods (the latter stochastic as well as deterministic) and for transferring the technology rapidly from one discipline to another. In particular, optimization schemes will be more effective when all available chemical information (e.g., function separability, availability of derivatives, additional experimental data) is taken into account in design of the algorithm, as is possible by preconditioning in both limited-memory quasi-Newton and truncated-Newton algorithms. Multigrid approaches (Kuruvila et al., 1994) and functional transformations (e.g., Piela et al., 1989; Wu, 1994) appear promising to global optimization problems in computational chemistry, and further developments might be fruitful.

Areas of mathematics that may have an important impact on the field are interval analysis and automatic differentiation. While the field of deterministic global optimization is still in its infancy in terms of general large-scale applicability, it is anticipated that the exploitation of vector and massively parallel computing environments for algorithm design will lead to significant progress in the coming years. Technological advances will clearly improve the range of global optimization strategies that can be considered, but greater efforts in parallel programming skills will be essential so that these high-performance platforms will have a true impact on these important scientific problems.

References

Bischof, C.H., A. Carle, G.F. Corliss, A. Griewank, and P. Hovland, 1992, ADIFOR: Generating derivative codes from FORTRAN programs, *Scientific Programming* 1:1–29.

Board, J.A., Jr., J.W. Causey, T.F. Leathrum, Jr., A. Windemuth, and K. Schulten, 1992, Accelerated molecular dynamics simulations with the parallel fast multiple algorithm, *Chem. Phys. Lett.* 198:89–94.

Byrd, R.H., C.L. Dert, A.H.G. Rinnooy Kan, and R.B. Schnabel, 1990, Concurrent stochastic methods for global optimization, *Math. Program.* 46:1–29.

Ciarlet, P.G., 1989, *Introduction to Numerical Linear Algebra and Optimization,* Cambridge University Press, Cambridge.

Dekkers, A., and E. Aarts, 1991, Global optimization and simulated annealing, *Math. Program.* 50:367–393.

Dembo, R.S., and T. Steihaug, 1983, Truncated-Newton algorithms for large-scale unconstrained optimization, *Math. Prog.* 26:190–212.

Dennis, Jr., J.E., and R.B. Schnabel, 1983, *Numerical Methods for Unconstrained Optimization and Nonlinear Equations,* Prentice-Hall, Englewood Cliffs, N.J.

Derreumaux, P., G. Zhang, B. Brooks, and T. Schlick, 1994, A truncated Newton minimizer adapted for CHARMM and biomolecular applications, *J. Comput. Chem.* 15:532–552.

Dixon, L.C.W., 1991, On the impact of automatic differentiation on the relative performance of parallel truncated Newton and variable metric algorithms, *SIAM J. Opt.* 1:475–486.

Dixon, L.C.W., and G.P. Szegö, 1975, *Towards Global Optimization*, Elsevier, New York.

Fletcher, R., 1987, *Practical Methods of Optimization,* Second Edition, John Wiley & Sons, New York.

Floudas, C.A., and P.M. Pardalos, eds., 1991, *Recent Advances in Global Optimization,* Princeton Series in Computer Science, Princeton University Press, Princeton, N.J.

Gilbert, J.C., and C. Lemaréchal, 1989, Some numerical experiments with variable-storage quasi-Newton algorithms, *Math. Prog.* 45:407–435.

Gill, P.E., W. Murray, M.A. Saunders, and M.H. Wright, 1983a, Computing forward-difference intervals for numerical optimization, *SIAM J. Sci. Stat. Comput.* 4:310–321.

Gill, P.E., W. Murray, and M.H. Wright, 1983b, *Practical Optimization*, Academic Press, N.Y.

Greengard, L., 1994, Fast algorithms for classical physics, *Science* 265:909–914.

Griewank, A., 1988, On automatic differentiation, in *Mathematical Programming 1988,* Kluwer Academic Publishers, Norwell, Mass. pp. 83–107.

Griewank, A., and G.F. Corliss, eds., 1991, *Automatic Differentiation of Algorithms: Theory, Implementation, and Application*, Society for Industrial and Applied Mathematics, Philadelphia, Pa.

Grubmüller, H., H. Heller, A. Windemuth, and K. Schulten, 1991, Generalized Verlet algorithm for efficient molecular dynamics simulations with long-range interactions, *Mol. Simul.* 6:121–142.

Hansen, E.R., 1980, Global optimization using interval analysis—The multidimensional case, *Numer. Math.* 34:247–270.

Hansen, E.R., 1992, *Global Optimization Using Interval Analysis*, Dekker, New York.

Hingerty, B.E., S. Figueroa, T.L. Hayden, and S. Broyde, 1989, Prediction of DNA structure from sequence: A buildup technique, *Biopolymers* 28:1195–1222.

Horst, R., and H. Tuy, 1993, *Global Optimization: Deterministic Approaches*, second edition, Springer, Berlin.

Kirkpatrick, S., C.D. Gelatt, Jr., and M.P. Vecchi, 1983, Optimization by simulated annealing, *Science* 220:671–680.

Kuruvila, G., S. Ta'asan, and M.D. Salas, 1994, Airfoil optimization by the *one shot* method, *Lecture Notes on Optimum Design Methods for Aerodynamics,* AGARD FDP/VKI Special Course, von Karman Institute for Fluid Dynamics, Rhode-Saint-Genese, Belgium, April, 25–29.

Leach, A.R., 1991, A survey of methods for searching the conformational space of small and medium-sized molecules, in *Reviews in Computational Chemistry*, Vol. II, K.B. Lipkowitz and D.B. Boyd, eds., VCH Publishers, New York.

Levy, A.V., and S. Gomez, 1985, The tunneling method applied to global optimization, in *Numerical Optimization 1984*, P.T. Boggs, R.H. Byrd, and R.B. Schnabel, eds., SIAM, Philadelphia, pp. 213–244.

Levy, A.V., and A. Montalvo, 1985, The tunneling algorithm for the global minimization of functions, *SIAM J. Sci. Stat. Comput.* 6:15–29.

Liu, D.C., and J. Nocedal, 1989, On the limited memory BFGS method for large scale optimization, *Math. Program.* 45:503–528.

Luenberger, D.G., 1984, *Linear and Nonlinear Programming*, Second Edition, Addison-Wesley, Reading, Mass.

Metropolis, N., A.W. Rosenbluth, M.N. Rosenbluth, A.H. Teller, and E. Teller, 1953, Equation of state calculations by fast computing machines, *J. Chem. Phys.* 21:1087–1092.

Nabors, K., F.T. Korsmeyer, F.T. Leighton, and J. White, 1994, Multipole accelerated preconditioned iterative methods for three-dimensional potential integral equations of the first kind, *SIAM J. Sci. Comput.* 15:713–735.

Nash, S.G., 1985, Solving nonlinear programming problems using truncated-Newton techniques in SIAM *Numerical Optimization 1984*, P.T. Boggs, R.H. Byrd, and R.B. Schnabel, eds., Philadelphia, pp. 119–136.

Nash, S.G., and J. Nocedal, 1991, A numerical study of the limited memory BFGS method and the truncated-Newton method for large-scale optimization, *SIAM J. Opt.* 1:358–372.

Nemhauser, G.L., A.H.G. Rinnooy Kan, and M.J. Todd, eds., 1989, *Handbook in Operations Research Management Science,* Vol. 1, Elsevier Science Publishers (North-Holland), Amsterdam.

Neumaier, A., 1990, *Interval Methods for Systems of Equations*, Cambridge University Press, Cambridge.

Nocedal, J., 1991, Theory of algorithms for unconstrained optimization, *Acta Numerica* 1:199–242.

Pardalos, P.M., and J.B. Rosen, 1987, *Constrained Global Optimization: Algorithms and Applications*, Lecture Notes in Computer Science 268, Springer, Berlin.

Piela, L., J. Kostrowicki, and H.A. Scheraga, 1989, The multiple-minima problem in conformational analysis of molecules. Deformation of the potential energy hypersurface by the diffusion equation method, *J. Phys. Chem.* 93:3339–3346.

Pincus, M., R. Klausner, and H.A. Scheraga, 1982, Calculation of the three-dimensional structure of the membrane bound portion of melittin from its amino acids, *Proc. Natl. Acad. Sci. USA* 79:5107–5110.

Purisima, E.O., and H.A. Scheraga, 1986, An approach to the multiple-minima problem by relaxing dimensionality, *Proc. Natl. Acad. Sci. USA* 83:2782–2786.

Rall, L.B., 1981, *Automatic Differentiation—Techniques and Applications*, Lecture Notes in Computer Science 120, Springer-Verlag, Berlin.

Rinnooy Kan, A.H.G., and G.T. Timmer, 1989, Global optimization, in *Handbooks in Operations Research and Management Science*, Vol. 1, G.L. Nemhauser, A.H.G. Rinnooy Kan, and M.J. Todd, eds., Elsevier Science Publishers (North-Holland), Amsterdam.

Scheraga, H.A., 1992, Predicting three-dimensional structures of oligopeptides, in *Reviews in Computational Chemistry,* Vol. III, K.B. Lipkowitz and D.B. Boyd, eds., VCH Publishers, New York, pp. 73–142.

Schlick, T., 1992, Optimization methods in computational chemistry, in *Reviews in Computational Chemistry*, Vol. III, K.B. Lipkowitz and D.B. Boyd, eds., VCH Publishers, New York, pp. 1–71.

Schlick, T., and A. Fogelson, 1992, TNPACK—A truncated Newton minimization package for large-scale problems: I. Algorithm and usage, and II. Implementation examples, *ACM Trans. Math. Softw.* 14:46–111.

Schlick, T., and M. Overton, 1987, A powerful truncated Newton method for potential energy minimization, *J. Comput. Chem.* 8:1025–1039.

Schnabel, R.B., 1989, Sequential and parallel methods for unconstrained optimization, in *Mathematical Programming*, M. Iri and K. Tanabe, eds., Kluwer Academic Publishers, Norwell, Mass., pp. 227–261.

Schnepper, C.A., and M.A. Stadtherr, 1993, Application of a parallel interval Newton/generalized bisection algorithm to equation-based chemical process flowsheeting, *Interval Computations* 4:40–64.

Shalloway, D., 1992, Application of the renormalization group to deterministic global minimization of molecular conformation energy functions, *J. Global Opt.* 2:281–311.

Törn, A., and A. Žilinskas, 1989, *Global Optimization,* Lecture Notes in Computer Science 350, Springer-Verlag, Berlin.

van Laarhoven, P.J.M., and E.H.L. Aarts, 1987, *Simulated Annealing: Theory and Applications*, D. Reidel, Dordrecht.

Wright, M.H., 1991, Interior methods for constrained optimization, *Acta Numerica* 1:341–407.

Wu, Z., 1994, *The Effective Energy Transformation Scheme as a Special Continuation Approach to Global Optimization with Application to Molecular Conformation,* Argonne National Laboratory report MCS-P442-0694, July.

Zou, X., I.M. Navon, F.X. Le Dimet, A. Nouailler, and T. Schlick, 1993, A comparison of efficient large-scale minimization algorithms for optimal control applications in meteorology, *SIAM J. Opt.* 3:582–608.

Locating Saddlepoints

The potential energy hypersurface of an individual molecule describes its minimum energy (i.e., stable) states as well as the transition structures linking these states. In other words, the local minima on the potential energy surface correspond to the minimum energy conformations of a molecule, and first-order saddlepoints on the surface correspond to transition states. These concepts can be extended to interacting molecular assemblies as well (e.g., clusters, biomolecular systems with solvent).

With the advent of modern computational techniques, it has become possible to exhaustively search the potential energy surface of individual molecules containing fewer than about 12 rotatable bonds (i.e., degrees of freedom for the dihedral angles that define molecular geometry, along with bond lengths and bond angles, in internal coordinate space) when classical (molecular mechanics-based) potential energy functions are employed. The previous section has broadly described the issues and various algorithmic techniques for finding local and global minima on such complex multidimensional energy surfaces. This section focuses on another aspect of conformational searches: the identification of saddlepoints and their connection to chemical reactions.

In addition to a description of the conformational properties of individual molecules, the potential energy surface can be employed to describe the energetics of chemical reactions. Therefore, searches on the potential energy hypersurface of a molecule can extend to molecular reactions as well (Eksterowicz and Houk, 1993). Reactants and products correspond to energy minima, whereas transition states linking products to reactants usually correspond to first-order saddlepoints on the energy surface (although unusual symmetries can produce higher-order transition states, including those of the "monkey-saddle" type). Thus, the location of stationary points (particularly minima and saddlepoints) on potential energy surfaces represents an important and challenging problem in computational chemistry.

In chemical applications, special conformational-space search methods have been devised for locating minima on molecular mechanics-based potential energy surfaces. These methods include stochastic (Saunders, 1987; Chang et al., 1989; Ferguson and Raber, 1989) and deterministic, grid-based (Motoc et al., 1986; Lipton and Still, 1988; Dammkoehler et al., 1989) approaches. Yet, with rare exception (Kolossvary and Guida, 1993), conformational searches have not been performed in such a way that saddlepoints are located. Nonetheless, the utility and indeed necessity of determining the conformational transition states that link these minima have recently been emphasized (Anet, 1990). Whereas in the past, conformational searches have been synonymous with location of energy minima, it is clear that in order to adequately study the conformational properties of molecules it is essential to locate first-order saddlepoints as well.

Significant effort has addressed the problem of locating transition states on potential energy surfaces derived from quantum mechanics calculations. A number of algorithms have been developed such as those that rely on eigenvector-following techniques (Cerjan and Miller, 1981; Simons et al., 1983, 1984; Bell and Crighton, 1984; Simons, 1985; Baker, 1986). In these methods one begins a saddlepoint search at or near a local minimum that is found by standard minimization techniques. A spectral decomposition is performed to find all the normal modes of the system (mass scaled eigenvalues and associated eigenvectors of the Hessian matrix (i.e., second-derivative matrix of the potential energy); then one of the normal modes is selected and followed in an "uphill" direction (i.e., a direction that leads to an increase of potential energy) until a saddlepoint is located. Evaluation of the energy gradient and Hessian matrix at each step of the search is performed until a point on the surface is located at which the gradient is zero and the Hessian possesses only one negative eigenvalue. In another approach, the linear synchronous transit method (Halgren and Lipscomb, 1977) has been employed to aid in the location of saddlepoints. It locates a maximum along a path connecting two structures and thus can be used to provide an initial guess for the transition state structure that connects them. Methods that find the location of saddlepoints by beginning the search at points on the potential energy surface that are of higher energy than the saddlepoint one wishes to locate have also been described (Berry et al., 1988). Recent developments (Jorgensen et al., 1988; Culot et al., 1992) have led to improved efficiency in locating transition states in calculations based on quantum mechanics-derived potential energy surfaces. Nonetheless, the aforementioned saddlepoint searches sometimes fail to converge, or they converge to critical points that are minima. Clearly, more robust algorithms are still needed, and this is an area that mathematical optimizers may find very interesting.

It is conceivable that algorithms for locating transition states on potential energy surfaces derived from calculations based on quantum mechanics could be employed for the location of conformational transition states on molecular mechanics-derived potential energy surfaces once the minima have been located. However, these algorithms have generally been used to study mechanisms of chemical reactions and have not been adequately tested for locating such conformational transition structures. In a typical conformational search procedure, the potential energy surface is scanned randomly or systematically and a large number of trial structures are generated for energy optimization. These structures can be severely "distorted" geometrically in the sense that bond lengths and angles lie out of the ranges observed experimentally, and van der Waals radii of atoms may overlap. These structures must then be optimized by the standard, "greedy" descent methods of local minimization toward a local minimum or toward a saddlepoint. However, for the quantum chemical calculation of a reaction mechanism, the reactant and product are usually known, and uphill movement toward the interconnecting saddlepoint is sought.

Conformational search procedures that locate first-order saddlepoints and minima with equal efficiency would be of enormous utility. Even though advances in this area have been slow, some progress has been achieved. For example, the so-called self-penalty walk method (Czerminski and Elber, 1990) provides an example of an algorithm for the calculation of reaction paths in complex molecular systems when molecular mechanics-derived potential energy functions are employed. However, it is likely that additional work will be required to develop methods for the efficient conformational searching of saddlepoints. New algorithms for conformational searches in which first-order saddlepoints are efficiently located are clearly urgently needed.

References

Anet, F.A.L., 1990, Inflection points and chaotic behavior in searching the conformational space of cyclononane, *J. Am. Chem. Soc.* 112:7172.

Baker, J., 1986, An algorithm for the location of transition states *J. Comput. Chem.* 7:385.

Banerjee, A., N. Adams, and J. Simons, 1985, Search for stationary points on surfaces, *J. Phys. Chem.* 89:52.

Bell, S., and J.S. Crighton, 1984, Locating transition states, *J. Chem. Phys.* 80:2464.

Berry, R.S., H.L. Davis, and T.L. Beck, 1988, Finding saddles on multidimensional potential surfaces, *Chem. Phys. Lett.* 147:13.

Cerjan, C.J., and W.H. Miller, 1981, On finding transition states, *J. Chem. Phys.* 75:2800.

Chang, G., W.C. Guida, and W.C. Still, An internal coordinate Monte Carlo method for searching conformational space, *J. Am. Chem. Soc.* 111:4379.

Culot, P., G. Dive, V.H. Nguyen, 1992, *Theor. Chim. Acta* 82:189.

Czerminski, R., and R. Elber, 1990, Self-avoiding walk between two fixed points as a tool to calculate reaction paths in large molecular systems, *Int. J. Quantum Chem.* 24:67.

Dammkoehler, R., S.F. Karasek, E.F.B. Shands, and G.R. Marshall, 1989, Constrained search of conformational hyperspace, *J. Computer-Aided Molec. Design* 3:3.

Eksterowicz, J.E., and K.N. Houk, 1993, Transition-state modeling with empirical force fields, *J. Chem. Rev.* 93:2439.

Ferguson, D.M., and D.J. Raber, 1989, A new approach to probing conformational space with molecular mechanics: Random incremental pulse search, *J. Am. Chem. Soc.* 111:4371.

Halgren, T.A., and W.N. Lipscomb, 1977, *Chemical Physics Letters* 49:1225.

Jorgensen, P., et al., 1988, *Theor. Chim. Acta* 73:55.

Kolossvary, I., and W.C. Guida, 1993, Comprehensive conformational analysis of the four- to twelve-membered ring cycloalkanes: Identification of the complete set of interconversion pathways on the MM2 potential energy hypersurface, *J. Am. Chem. Soc.* 115:2107.

Lipton, M., and W.C. Still, 1988, The multiple minimum problem in molecular modeling. Tree searching internal coordinate conformational space, *J. Comput. Chem.* 9:343.

Motoc, I., R.A. Dammkoehler, D. Mayer, and J. Labanowskki, 1986, Three-dimensional quantitative structure-activity relationships. I. General approach to the pharmacophore model validation, *Quant. Struct.-Act. Relat.* 5:99.

O'Neal, D., J. Simons, and H. Taylor, 1984, Potential surface walking and reaction paths for $C_2VBe + H_2$)-Be + $H_2 \rightarrow$ Be + 2H (1A1), *J. Phys. Chem.* 88:1510.

Saunders, M., 1987, *J. Am. Chem. Soc.* 109:3150.

Simons, J., P. Jorgensen, J. Ozment, H. Taylor, et al., 1983, Walking on potential-energy surfaces, *J. Phys. Chem.* 87:2745–2753.

Simons et al., 1984, *J. Phys. Chem.* 88:1519.

Simons, J., N. Adams, A. Banerjee, and R. Shepard, 1985, Search for stationary-points on surface, *J. Phys. Chem.* 89:52–57.

Sampling of Minima and Saddlepoints

Many problems in computational chemistry require a concise description of the large-scale geometry and topology of a high-dimensional potential surface. Usually, such a compact description will be statistical, and many questions arise as to the appropriate ways of characterizing such a surface. Often such concise descriptions are not what is sought; rather, one seeks a way of fairly sampling the surface and uncovering a few representative examples of situations on the surface that are relevant to the appropriate chemistry. Some specific examples include finding snapshots of crucial or typical configurations or movies of kinetic pathways. This allows what one might call an artistic description of the chemical situation. Such a description is often looked down upon by quantitative scientists as being "anecdotal," but it is important not to cut ourselves off from any route to understanding. To make this point one might compare the kinds of understanding of ancient cultures that are obtained from the numerous scholarly statistical studies of bookkeeping accounts and what we learn from the great paintings of the same periods, which give us different perspectives on social life. The main danger of such artistic representations is that one must have some guarantee that they do not simply represent a kind of beautiful propaganda for an incorrect qualitative viewpoint. Clearly, statistics must be used to validate such individual samples of a system's behavior.

Several chemical problems truly demand the solution of these mathematical problems connected with the geometry of the potential surface. Such a global understanding is needed to be able to picture long time scale complex events in chemical systems. One area in which this is clearly essential is the understanding of conformational transitions of biological molecules. The regulation of biological molecules is quite precise and relies on sometimes rather complicated motions of a biological molecule. The most well studied of these is the so-called allosteric transition in hemoglobin, but indeed, the regulation of most genes also relies on these phenomena. These regulation events involve rather long time scales from the molecular viewpoint. Their understanding requires navigating through the complete configuration space. Another such long time scale process that involves complex organization in the configuration space is biomolecular folding itself. By what process is the structure of a biological molecule determined? In order to function, enzymes require a fairly precise three-dimensional positioning of different chemical groups in the protein molecule. To achieve this precise positioning of only a few groups, the collective interactions of the rest of the molecule must conspire to form such a fairly rigid construction. Although the three-dimensional structures of protein molecules exhibit some symmetries, they are exquisitely complex, and in addition, the architectures of folded protein are formed from molecules that have no simple pattern in their one-dimensional sequence.

Understanding how the Brownian motion on an energy surface can funnel such a molecule to a very precise structure is a major puzzle requiring a global analysis of the many-dimensional energy surface. The global geometry of the potential energy surface also enters into the study of nonbiological chemical problems such as those involving the structure and mechanical properties of amorphous materials. While crystalline solids can be studied through the analysis of the ground state and the first few excited states, glasses and other amorphous materials have a huge number of local structural configurations. Unlike a typical liquid, however, these individual configurations last for incredibly long periods of time, and one must understand the statistics of the different minimal energy structures and the nature of the transitions between them in order to quantify the slow relaxations of such systems. The aging of amorphous materials in glasses shows that they do not obey the simple equilibrium statistical mechanical laws so often used to characterize simple materials. At the same time, these aging phenomena have great practical significance to the macroscopic properties (e.g., long time stability) of materials important to such applications as fiber optics. The local minima are the mathematically simplest objects to characterize statistically the potential energy surface. The crucial questions here are first, what do the minima look like and, second, how many of the different kinds of minima exist on the surface? Sampling many minima of a potential energy surface can be carried out with gradient descent techniques, and a great deal has been learned about the qualitative structural characteristics of these minima for both biomolecules and glasses. The counting of minima seems to be of crucial importance as well, since at the phenomenological level, the kinetics of amorphous materials are highly correlated with their configurational entropy. At this point, no very good algorithm yet exists for doing this sort of counting in an objective and reliable way, even on a computer.

Statistical mechanics is the study of the collective behavior of large numbers of interacting particles. Properties of interest include those describing time-dependent, irreversible process. The basic principles of this discipline were laid down in the nineteenth century by Ludwig Boltzmann, James Clerk Maxwell, and Josiah Willard Gibbs.

The very deepest minima of systems can be characterized by using techniques superficially similar to those of thermodynamics and equilibrium statistical mechanics. Generalization of mean field theory for random Hamiltonians is used. The low-lying states of heteropolymeric biological macromolecules have been studied in this way. There is a very clear analogy to the phenomenon of broken ergodicity studied in spin glasses by the quasi-equilibrium statistical mechanical methods. The problem of broken ergodicity is one that is central to understanding the global topology of potential energy surfaces for such "random" systems. This problem plays a role both in the issues discussed here of biological macromolecules and amorphous materials, and in other optimization problems as well. There are deep connections with the theory of NP-completeness, a fundamental question in theoretical computer science. The formal questions of broken ergodicity in spin glasses (i.e., the topology of low-energy states) have not been answered entirely unambiguously by experiment, and the question of the nature of the low-lying states is one that is still hotly debated. An important route to understanding this sort of broken ergodicity has been by the methods of rigorous statistical mechanics pioneered by mathematicians. It has been shown rigorously in some higher-dimensional problems that the broken ergodicity imagined in simple phenomenological theories of protein folding can, in fact, occur. It is still an open question, however, how ergodicity is broken for three-dimensional systems, spin glass systems, or for the random heteropolymers themselves.

Ergodicity is the capacity of a dynamical system spontaneously to sample all of its phase space.

One of the most interesting results of the theory of broken ergodicity based on quasi-equilibrium statistical mechanics is that the low-energy states of a typical Hamiltonian are related to each other in a fashion that is characterized by an ultrametric distance. This ultrametricity concept arose earlier in the study in pure mathematics. The ultrametric organization may well play a role in the dynamics on such surfaces, and ultrametric hopping models have been widely discussed.

While the use of statistical energy surface topography is now coming to be accepted in the context of biomolecules, there is a still deeper mathematical question in its application to glasses. This question is, How does a Hamiltonian that is perfectly regular, having no explicit randomness, possess solutions that appear to be totally irregular and aperiodic? A long-standing issue for the purist has been whether even hard spheres have, as their most dense state, the simple regular packing characteristic of face-centered cubic (FCC) crystals. Recently a proof of this was announced, but it has apparently been retracted. In fact, for the three-dimensional situation there is little doubt from the experimental input that the dense state is in fact periodic. The question of the closest packings in

BOX 4.6 Comments on the Ambiguous Concept of "Structure" for Complex Molecules and Macromolecules

Owing to the special nature of their roles in life processes, biological macromolecules such as proteins, nucleic acids, and carbohydrates have evolved into systems exhibiting a high degree of structural diversity and complexity. Theoretical/computational chemistry bears the responsibility of predicting, characterizing, and explaining the biological reasons for the three-dimensional shapes or "structures" adopted by those macromolecules.

However, researchers should be aware (particularly on first entering this problem area) that the concept of "structure" has multiple interpretations that depend on the physical and chemical circumstances involved.

As explained in the main text, prediction of the preferred structure adopted by any given molecule is usually reduced in principle, if not in practice, to the study of minima on a suitable energy surface in a multidimensional space of configurational coordinates. In many—but not all—cases, the global minimum corresponds to the biologically active structure, while higher-lying relative minima correspond to inactive denatured forms.

The figure below shows, in simple cartoon fashion, three generic energy surfaces. The simplest (a) contains but a single minimum that would be easy to locate numerically. The next (b) shows multiple minima and requires more effort if a full classification of extrema is warranted by the problem it represents. Case (c) is most representative of the situation with biological macromolecules, with a vast array of minima arranged in basins or valleys over a wide range of length scales. In this last circumstance, the concept of "structure" depends in part on the level of accuracy that is warranted, and that level is strongly dependent on temperature.

At very low temperature (e.g., a protein frozen in its aqueous medium), thermally excited vibrations will be so feeble that the system of protein and water molecules would be trapped in the vicinity of a single fine-grained minimum. Raising the temperature stimulates transitions between neighboring microbasins, so the relevant notion of "structure" entails the average configuration for the broadened distribution. A simple view would be that raising the temperature effectively smooths out the finer features of the complicated energy surface. This amounts to passage from (c) to (b) in the figure.

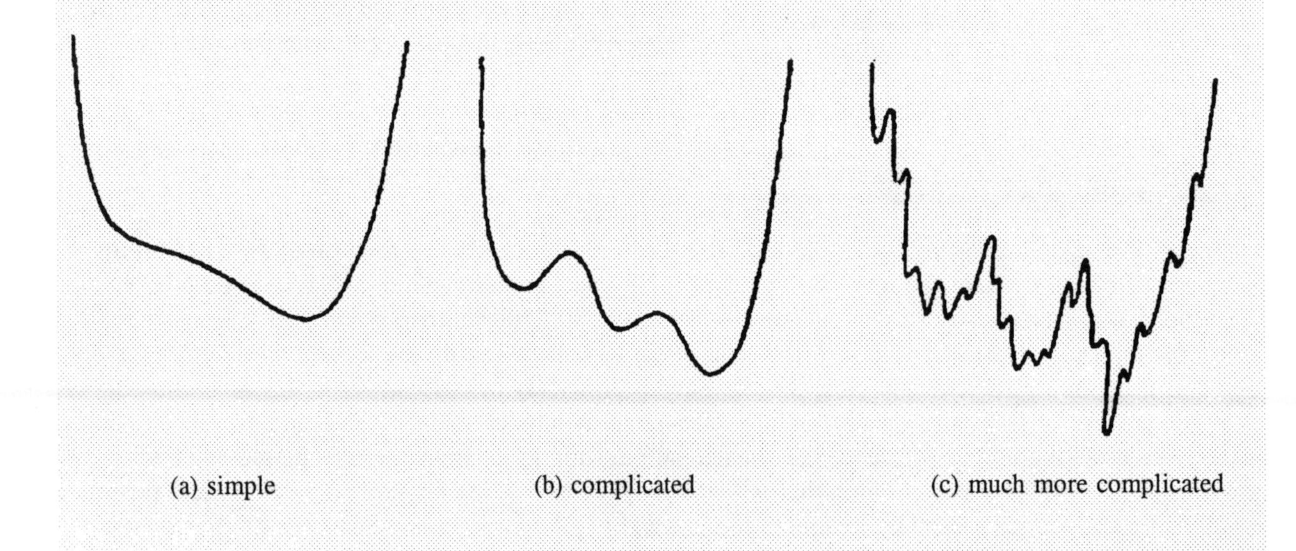

In the vicinity of room temperature, where the aqueous medium is liquid, it is traditional to average over the solvent degrees of freedom and to utilize the resulting "free energy surface" for biological macromolecule studies. Consequently, the configuration space undergoes a reduction in dimensionality to that of just the macromolecule's flexible degrees of freedom. At the same time, the surface to be searched for minima becomes temperature dependent, so the number of minima it exhibits, where they are located, and indeed which is the absolute minimum can vary. Furthermore, interbasin transitions can produce substantial configurational fluctuations even in this reduced-dimensionality representation, requiring a correspondingly permissive definition of "structure."

high-dimensional systems has many contacts with group theory and the theory of optimal coding.

The existence of quasicrystals has made the problem an even richer one since even the existence of such quasi-periodic structures was ruled out by "folk theorems" of physicists. It is likely that truly aperiodic crystals can, in principle, exist in three dimensions. An important argument for this is based on tiling theory. It has been shown that certain tiling problems are NP-complete. This implies that it is certainly very difficult to figure out whether a periodic packing of such a tiling is possible. Thus, it seems that the problem of totally aperiodic crystal phases for regular structures may be itself tied to the NP-completeness question.

The problem of transitions between minima on such a high-dimensional surface is in a still more primitive state than the characterization of minima. The search for minima is itself a relatively stable computational problem. The search for saddlepoints that connect individual minima is computationally much more difficult. This is certainly a consequence of the unstable mode at such saddlepoints. Despite numerous efforts, there are no entirely reliable methods for carrying out such a search. For many simple problems, finding a reasonably good transition state is possible, but these techniques become still more complicated and less reliable as system size increases. On the purely theoretical side, very simple models that relate the heights of barriers to the statistics of minima have been developed, but almost no truly rigorous work has been done. Simply characterizing the minima and the saddlepoints connecting a few of them does not give an entire description of significant processes on a complex energy landscape. It is clear that one must understand something more about the basin of attraction of any given minimum. If the nearby minima are not entirely uncorrelated, this basin of attraction will depend on their structure as well. A characterization of the size of such funnels in biomolecular problems is essential to understanding protein folding.

Similarly, in many such complex problems it has been imagined that specific kinetic pathways are important. Again some work has already been done on the question of how specific pathways can emerge on a statistical energy landscape. These ideas are, however, based on the quasi-equilibrium statistical mechanics of such systems, and there are many questions about the rigor of this approach. Similarly, a good deal of work has been carried out to characterize computationally pathways on complicated realistic potential energy surfaces.

Techniques based on path integrals have been used to good effect by Elber in studying the recombination of ligands in biomolecules and in the folding events involved in the formation of a small helix from a coiled polypeptide. These techniques tend to focus on individual optimal pathways, but it is also clear that sets of pathways are very important in such problems. How these pathways are related to each other and how to discover them and count them is still an open computational challenge.

BOX 4.7 Implications of Dynamical Chaos at the Classical Level

A large part of modern computational chemistry is based on the solution of the equations of classical mechanics for many-body systems. For these problems, the standard numerical integration techniques found in classic textbooks only provide a simple framework for application and analysis. This is because most of the theory of numerical analysis provides criteria for long-time stability for these smaller systems with regular dynamics. Nevertheless, it is clear that for *most* many-body *chemical* systems, the differential equations may have chaotic solutions primarily. Thus, it is of little use to talk about the stability of an individual trajectory for long times, particularly when methods with some stochastic elements (e.g., Langevin dynamics) are involved.

A necessary mathematical advance is an understanding of error estimation and long-time stability for chaotic systems. In fact, one seeks a way of characterizing the accuracy in some statistical sense from such a simulation, or collection of simulations, since an individual trajectory's *details* are certainly predicted incorrectly. The classical theory of error estimates in numerical analysis is clearly inadequate for most purposes and the theory of long-time stability for complex systems is rather at its infancy (Stuart and Humphries, 1994).

A most striking example is in the old simulations of hard-sphere molecular dynamics showing trajectories for times encompassing many, many collisions of a dilute gas (e.g., Alder and Wainwright, 1970). One can easily show that the trajectories were numerically inaccurate beyond the limit of machine precision, even after only 10 collisions. Nevertheless, when the velocity correlation function was computed, it exhibited a long-time tail that persisted out to 30 collision times, and this long-time tail's form as well as amplitude agreed precisely with kinetic theory calculations. Thus, statistical properties can be accurate even when individual trajectories are entirely incorrect.

A similar problem enters when one considers the use of "stochastic dynamics" methods in simulation (e.g., pseudorandom forces are added to the systematic force to mimic a thermal reservoir). When only a subsystem of a larger system (e.g., biomolecule plus solvent) is being studied, the appropriate equations of motion are stochastic or Langevin equations. The most familiar example is the diffusion of biomolecules in water. The theory of quadratures for such stochastic equations has long been of interest, but comprehensive analyses of associated error estimates have been developed mainly from a relatively simple point of view, such as via analysis of harmonic oscillators (Pastor et al., 1988; Tuckerman and Berne, 1991).

However, there are many important mathematical issues that arise in this connection. What is the meaning of "error analysis" in this stochastic framework? Is it appropriate to compare results to parallel approaches, such as molecular dynamics, where the stochastic forces are zero? Are asymptotic results as the time step approaches zero relevant to practical problems? As discussed elsewhere in this report, the complexity of biomolecular systems involves multiple conformations and numerous biologically relevant pathways. And, given the severity of the time step problem in molecular dynamics (see pages 54–58), how can qualitative and quantitative theories for evaluating simulation results be merged? Clearly, there is a strong need for both aspects, but one can imagine that different numerical models in combination with different integration or propagation methods could be designed to address various aspects of the dynamical problems. Therefore, an organized theory for simulation evaluation, including local error analysis, long-time behavior, and some kind of broader or "global" framework for evaluation, would be extremely valuable. Such a theory could clearly aid in evaluating new models and methods as they arise, identifying the appropriate tests for the various simulation protocols, and putting in perspective the biological results that may emerge from any computer simulation.

References

Alder, B.J., and T.E. Wainright, 1970, *Phys. Rev. A* 1:18–21.

Pastor, R.W., B.R. Brooks, and A. Szabo, 1988, An analysis of the accuracy of Langevin and molecular dynamics algorithms, *Mol. Phys.* 65:1409–1419.

Stuart, A.M., and A.R. Humphries, 1994, Model problems in numerical stability theory for initial value problems, *SIAM Review* 36:226–2571.

Tuckerman, M., and B.J. Berne, 1991, *J. Chem. Phys.* 95:4389.

Efficient Generation of Points That Satisfy Physical Constraints in a Many-Particle System

Prototypical Problem

Consider N particles in a cube in three-dimensional space, each with x, y, and z coordinates in the range $[0, L]$. The state of the system is then given by specifying the positions r_i of all N points.

In the statistical mechanics modeling of condensed phases, one typically is interested in restricted sets of particle configurations. For instance, one's interest is often restricted to only those states for which

$$|\boldsymbol{r}_i - \boldsymbol{r}_j| \geq 1 \quad \text{for all } 1 \leq i < j \leq N .$$

Imagine that the points represent the locations of the centers of hard spheres of diameter unity. The conditions state that the hard spheres do not overlap one another; that is, the spheres repel each other strongly when they are close together, so that each center-center distance must be greater than unity.

For the present problem, we are interested in the regime where N is large, of the order of 10^2 to 10^6. The volume is large enough so that N/L^3 is in the range $(0,2^{1/2})$. In the actual problem of common interest for computer simulation of materials, the system will satisfy periodic boundary conditions. In effect, this means that the restriction above is more precisely stated as

$$|\boldsymbol{r}_i - \boldsymbol{r}_j - \boldsymbol{n}L| \geq 1 \quad \text{for all } 1 \leq i < j \leq N$$
$$\text{and all vectors } \boldsymbol{n} \text{ with integer components} .$$

The problem is to generate efficiently states of the system that satisfy these constraints. The states will more than likely be generated by a stochastic process of some sort. A more ambitious goal is to generate states such that all states that satisfy the constraints are equally likely to be generated. (More precisely, if the set of positions $\{r_1, \ldots, r_N\}$ is regarded as a set of random variables, the joint distribution function for these variables, which is zero when one or more of the constraints is violated, should be a constant for values that satisfy the constraint.)

This is an example of a problem for which each constraint is relatively easy to state and express in terms of a small number of the variables in the problem, and the number (or measure) of states that are consistent with the constraints is very small compared with the total number of states, in fact vanishing exponentially as N increases. This problem is related to that of generating possible states for an atomic fluid whose interatomic potential precludes two atoms from getting close to one another. The rigid-sphere interpretation relates to the challenging mathematical problem of existence and characterization of random sphere packings.

Variations on the Prototypical Problem

First Variation. Consider a random walk in a three-dimensional space that consists of N steps of unit length and random direction. Let s_i for $i \geq 1$ be the ith step. The position

$$\boldsymbol{r}_i \equiv \sum_{j=1}^{i} \boldsymbol{s}_i$$

is the location of the random walker after the ith step, with r_0 being equal to the origin. The only states of interest are those for which these positions never come close to one another. More precisely,

$$|\boldsymbol{r}_i - \boldsymbol{r}_j| \geq 1 \quad \text{for all } 0 \leq i < j \leq N.$$

This problem is related to specifying the possible structures of a polymeric molecule. The goal is to generate random walks that satisfy these conditions. As in the previous problem, the method of generation is likely to be probabilistic. A stronger goal would be to generate states with a process such that all states that satisfy the constraint are equally likely to be generated. (More precisely, if each step $\boldsymbol{s}_i$ of the random walk is specified by a location on the surface of a unit sphere, the joint distribution function of the N steps should be zero for sets of steps that violate the constraints and a constant for sets that satisfy the constraints.)

Second Variation. In problems of interest, there may also be some additional constraints on the locations; for example, for some pairs of locations, there might be restrictions of the form

$$a_{ij} \leq |\boldsymbol{r}_i - \boldsymbol{r}_j| \leq b_{ij}.$$

When the number of these restrictions is large enough to imply that only a small range of structures can satisfy all the constraints, the problem becomes a special case of the problems that are solved by using the methods of distance geometry, which is discussed in more detail in Chapter 3.

Third Variation. The various steps in the random walk of the previous problem might be correlated in the sense that the probability distribution for $\boldsymbol{s}_i$ might depend on the value of $\boldsymbol{s}_{i-1}$ or perhaps both $\boldsymbol{s}_{i-1}$ and $\boldsymbol{s}_{i-2}$.

In these problems, there are some constraints that are "easy" to satisfy (e.g., in the first problem each particle must be inside the cubic box; in the second problem, each step must have unit length). Then there are others that are "harder" to deal with. Whether a constraint is hard or easy to deal with is related, in general, to whether it is concerned with just one of the basic vectors of the problem or with more than one.

Simplest Strategy

The simplest strategy for generating states that satisfy all the constraints is obviously to generate states that satisfy the easy constraints and then delete those states that violate the hard constraints. This solution is practical, if at all, only for relatively uninteresting situations. It works for the first problem, for example, only when the density N/L^3 is much lower than the maximum density allowed. The problem is that almost all of the states generated will subsequently be deleted by this process.

More precisely, in the first problem one might imagine generating sets of N points, each of which is randomly distributed in the cube, and then discarding sets that violate the conditions specified. The sets not discarded then should be uniformly and randomly distributed among the states that do satisfy the conditions. The difficulty with this approach is that the probability that a set of randomly generated points satisfies the condition is of order $\exp(-aN)$ for large N, where a is a constant that depends on the density N/L^3. For large N, therefore, much of the computational effort is wasted.

An obviously more efficient procedure is to generate the set of positions one at a time and test each one to ensure that it is far enough from the previous positions before generating the next position. If one violation of the conditions is found this way, no effort need be expended to generate the remaining positions in the set or to test them. Even this is not efficient enough. A significant amount of effort will be expended in generating partial sets of positions, only to find that the set must be discarded because of the value of some position generated late in the sequence. This difficulty is sometimes referred to as the problem of "exponential attrition" because of the exponentially small fraction of sets of positions that are generated successfully.

Metropolis Monte Carlo Method

In this strategy one first generates one state that satisfies all the constraints; then this is used as the beginning of a Markov process whose transition probabilities are such that transitions are allowed only to other states that satisfy the constraints. In practice, this means that each transition typically involves a change of, at most, one of the coordinates by a very small amount. The two difficulties with this method are the following:

1. The set of states that satisfy the constraints may not be connected, so that with a particular initial state it will be impossible to generate a very large fraction of the states.

2. Even if the set of states that satisfy the constraints is connected, typical Markov processes explore the range of accessible states relatively slowly.

Therefore, some entirely new ideas for dealing with this class of problems would be worthwhile.

Relationship of These Problems to More General Optimization Problems

Some special cases of these problems, especially the first and second variations described above, are closely related to optimization problems that arise in chemical calculations. Chemical optimization problems typically involve minimization of an energy or free energy that depends on the positions of atoms or groups of atoms. (See pages 68–77 for a discussion of optimization problems and methods.) In more general problems, the object is not to minimize the energy or free energy but to calculate the typical properties of all the states of low enough energy that they might be populated at the temperature of interest. Such functions typically are very large and positive for certain configurations in which atoms or molecules are very close. It is typically true that a configuration in which any one pair of atoms is too close to one another has a high enough energy to make the total energy of the configuration so high that it cannot possibly be a solution of the optimization problem (or so high that it is not thermally populated). Thus, identification of states that are consistent with constraints of the type mentioned can be a useful first step toward solving optimization problems in chemistry.

Molecular Diversity and Combinatorial Chemistry in Drug Discovery

Overview of the Drug Discovery Process

The discovery of new drugs is a time-consuming, risky, and expensive process. These things are true even though in the past 15 years there has been a dramatic increase in the number of three-dimensional structures of proteins that can be used as scaffolds for the conceptual and computational aspects of drug design. The discovery traditionally moves through several stages once the biological target has been chosen (see *Science*, 1994).

First, a moderately active compound, a "lead," is identified from clues provided by the literature, through "random" screening of many compounds or through targeted screening of compounds identified by three-dimensional searching or docking. If the three-dimensional structure of the biological target macromolecule is known, then one may use three-dimensional searching to identify existing compounds that are complementary to the binding site in the target (Kuntz, 1992; Martin, 1992). Alternatively, if a number of structurally unique compounds bind to the target, one may propose a three-dimensional pharmacophore and search databases for matches to it (Martin, 1992).

Once a lead has been identified, hundreds to thousands of additional compounds are designed and synthesized to optimize the biological profile. The cost and environmental impact of synthesis and patentability are also important issues. If the three-dimensional structure of the biological target is known, then molecular modeling might be used in the design (Erickson and Fesik, 1992). As testing data are accumulated, statistical three-dimensional quantitative structure-activity relationships (three-dimensional QSAR) may help set priorities for synthesis. Any attractive compounds found are tested in more detail through advanced protocols that more reliably forecast therapeutic and toxicity potential.

> A **pharmacophore** is a chemical identity and geometrical arrangement of the key substituents of a molecule that confer biochemical or pharmacological effects.

Lastly, the surviving compounds are prioritized, and the best compound known at that time is prepared for clinical trial.

New mathematical techniques could have an impact on the rate of new compound discovery if the potency of compounds could be forecast more quickly and accurately before their synthesis. Many of the improvements in computational chemistry discussed elsewhere in this report would also impact the ability to forecast affinity based on the structure of the ligand and the macromolecular target. However, additional opportunities exist for cases in which the structure of the macromolecular target is not known, cases for which the forecast is based on three-dimensional QSAR investigations (Kubinyi, 1993). The most explored method is comparative molecular field analysis (CoMFA; Cramer et al., 1988). With CoMFA, molecules are aligned with each other; then for each molecule, the interaction energies with various probes are calculated at intersections of a three-dimensional lattice that encloses all the molecules. The relationships between these thousands of energy values and the potencies of the 10 to 100 molecules are established by the statistical method of partial least squares (PLS) with leave-one-out cross-validation (see Frank and Friedman, 1993, for background on PLS and comparisons of the method to other statistical procedures). When a CoMFA model is found, it generally has quite robust forecasting ability: the average error in forecasting the potency of 85 compounds in eight datasets is 0.55 logs or 0.8 kilocalories per mole (Martin et al., in press).

However, there are indications that one may fail to find a model, even though one exists, because of the coarseness of the lattice spacing (2 Å) and the sensitivity of PLS to noise. PLS can find only linear relationships between properties and biological potency; a method that could detect nonlinear relationships would be an improvement and might model more sets of data. Limited experiences with neural nets have shown no improvement over PLS. There might be an optimization method that could select the relevant variables from a pool of thousands. It would have to be roughly as fast as PLS (a minute or so to do leave-one-out cross-validation on 25 compounds) since one of the elements of the analysis is to compare results with different properties calculated at the lattice points, adding whole molecule properties, comparing alignment rules, investigating outliers, and combining and separating subseries of molecules.

Sources of Molecular Diversity

The weak point in the whole scenario of new drug discovery has been identification of the "lead." There may not be a "good" lead in a company's collection. The wrong choice can doom a project to never finding compounds that merit advanced testing. Using only literature data to derive the lead may mean that the company abandons the project because it cannot patent the compounds found. These concerns have led the industry to focus on the importance of molecular diversity as a key ingredient in the search for a lead. Compared to just 10 years ago, orders of magnitude more compounds can be designed, synthesized, and tested with newly developed strategies. These changes present an opportunity for the imaginative application of mathematics.

Automated testing methods employing simplified assays, mixing strategies, robotics, bar-coding, etc., have led many pharmaceutical and biotechnology companies to test every available compound, perhaps 10^5 to 10^6 of them, in biological assays of interest (Gallop et al., 1994; Gordon et al., 1994). Testing a collection generally takes approximately six months. This operation presents several challenges: (1) Is it really necessary to test all of the compounds in order to identify the series of compounds that will show the activity? (2) Should a pilot set of compounds be tested first to adjust the assay conditions and forecast how many active compounds will be found? If so, how would this set be selected? (3) What compounds, available from outside vendors, should be selected for purchase to complement the set of in-house compounds? Is there a way to quantify their worth other than the cost to synthesize in-house?

Concurrently, synthetic chemists developed new strategies that provide large numbers of compounds for biological testing typically as mixtures. Such libraries, synthesized in a few months, can contain 10^4 to 10^7 different chemical structures (Baum, 1994). Although this number of compounds seems high, note that it has been estimated that there are 10^{200} stable chemical compounds of molecular weight less than 750 that contain only carbon, hydrogen, nitrogen, oxygen, and sulfur. Even factoring in their possibility of synthesis and realistic chemical and physical properties still leaves on the order of 10^{180} compounds to consider. How, then, does one choose which 10^4 compounds should be included in the first library, or the second?

A final strategy to enhance molecular diversity results from computer programs that design molecules to meet specified three-dimensional criteria, typically based on the experimental structure of a protein binding site (Rothstein and Murcko, 1993). The programs design molecules to meet geometric criteria and include electrostatic complementarity at the level of force fields such as those used for molecular dynamics. The diversity arises from the combinatorics: a protein binding site usually contains at least four or five hydrogen-bonding or charged groups; a ligand might interact with most or all of them, and many different templates might be able to fit into the binding site and orient polar groups for optimal interaction. Hence, it is expected that a huge number of nicely fitting molecules might be designed. Although design programs could be set up to produce only those molecules that could be synthesized readily, this severely limits the diversity. Hence, it is likely that the designed molecules will have to be made by traditional synthesis. This places a realistic upper limit of 25 molecules to be selected. Even if binding affinity could be forecast precisely, we are a long way from forecasting every type of toxicity or drug metabolism quirk that a molecule might possess. Again, we face the problem of selection of the most diverse sample from a population.

Current Computational Approaches to Compound Selection

There are three aspects to the problem of selecting samples from large collections of molecules: First, what molecular properties will be used to describe the compounds? Second, how will the similarity of these properties between pairs of molecules be quantified? Third, how will the molecules be grouped or clustered?

For datasets of size 10^4 and higher, the standard method of describing the molecules for clustering encodes the presence or absence of substructural features in a bit-string, typically of length 256–1024 (Willett, 1987; Hodes, 1989). In modern systems, these substructural features are recognized by enumerating all paths of length 0–7 in the molecular graph and using these to populate one or more of the bits (Weininger et al., 1994). It typically takes one to two hours on a modern workstation to generate such *fingerprints* of a database of 10^5 compounds. The time required for this process increases linearly with the number of compounds.

BOX 4.8 Possibility of Intelligent Algorithms to Detect Novel Phenomena Automatically

The total amount of numerical information generated in a typical computer simulation is immense. Whether the model system under study is inorganic or biological, classical or intrinsically quantum mechanical, in thermal equilibrium or violently nonstationary, the processing of the data created is usually structured to address a small set of often conceptually orthodox questions that have been selected beforehand. Given the inappropriateness of trying to examine raw numerical information in tabular form, this is entirely reasonable. Nevertheless, serendipitous discoveries that fall outside the preselected query set indeed do occasionally occur, and their consequences can produce significant advances.

The usual procedures for carrying out computer simulations probably are far from optimal in detecting initially unsuspected correlations and phenomena. This leads one to suspect that generic pattern-recognizing algorithms might be constructed that would operate in conjunction with Monte Carlo or molecular dynamics simulations and would effectively amplify the human scientist's limited capacity to pull novel correlations out of huge files of numerical data. Generally speaking, humans are best at recognizing visual patterns—after all, the human visual apparatus has been honed to exquisite sensitivity by approximately a billion years of evolutionary trial. But in the scientific context this requires that just the right graphs or figures be presented to the investigator, and in one respect it presupposes something about the nature of the correlations to be examined. One hopes to transcend such human limitations, perhaps with a carefully crafted form of open-ended artificial intelligence that incorporates pattern-detecting capacity free from our human biases and limitations. In particular, computers ought to be able to "see" in spaces of dimension higher than three and pick out significant patterns whose projection into two or three dimensions would obscure the phenomena of interest.

This opportunity is not so much intended as an attempt to replace human researchers as it is to magnify their wisdom and insight. In this respect, generic pattern-detecting software could be a powerful device operating to maximize productivity. And it should not escape notice that such a tool is not restricted just to simulation data, but in principle could also apply to other classes of large chemical databases discussed elsewhere in this report.

The second step is to calculate the similarity of every molecule to every other molecule in the dataset. The similarity measure traditionally used, the Tanimoto coefficient, is expressed as

$$\mathrm{Sim}_{ij} = \frac{F_{ij}}{F_i + F_j + F_{ij}},$$

where Sim_{ij} is the similarity of molecule i to molecule j, F_{ij} is the number of features (bits set to 0 or 1) in common between molecule i and molecule j, F_i is the number of bits set in molecule i, and F_j is the number of bits set in molecule j. For the same 10^5 compounds, this process takes on the order of 24 hours. Since every molecule is compared with every other, it scales as the square of the number of compounds. Lastly, the Jarvis-Patrick clustering method (Jarvis and Patrick, 1973) is used to group the compounds. This method is based on comparing the nearest neighbors of compounds and is very fast, taking only seconds to accomplish. Although each of these steps is feasible, none is optimal.

Opportunities for Improvements in Computational Approaches to Compound Selection

Molecular fingerprints are not the best descriptor to use to select compounds for bioactivity since the biological properties of compounds depend on their three-dimensional complementarity of shape and electronic properties with those of the target biomolecule. Clearly, we would like to consider the three-dimensional structures of the molecules—shape, the location of intermolecular recognition sites such as hydrogen-bonding or charged groups, and the way in which the position of these features changes with changes in conformation. Speeding up such calculations or representations of the results would be a big help.

However, the problem of how to represent conformational flexibility in this context is a bigger challenge—do we need a totally different way to represent the structures than coordinates or distances between pairs of atoms? It is important to recognize that two-dimensional molecular structure is the basis of both chemical synthesis strategy and patent claims, and so the representation must also include the two-dimensional structure. This is a clear example where the choice of model is significant and must be the result of close collaboration between mathematical and chemical scientists.

Any new molecular descriptor will require that one define a corresponding metric for the similarity or distance between compounds to be used in grouping them. For example, in contrast to substructural features, which are either present or absent, distances are continuous and do not fall so easily into a bit-string. Should distances be binned—if so, should the bins be fuzzy or overlapping? How is similarity evaluated in such cases?

On the other hand, one might quantitate the similarity of two molecules by the size and composition of the maximum common substructure. Experience with the Bron-Kerbosh algorithm (Bron and Kerbosh, 1973) has shown a rate of 10^6 comparisons per hour. For a dataset of size 10^5, it would take 10^{10} comparisons or 10^4 hours to prepare the similarity matrix! Similarly, 10^6 molecules would require 10^{12} comparisons and 10^6 hours. Although parallelization might allow one to perform the calculations, a better algorithm might accomplish the same thing with greater efficiency. It might be possible to eliminate most of the comparisons while retaining all important ones.

Improvements in the grouping of compounds are sorely needed. The Jarvis-Patrick algorithm performs poorly on sets of very diverse compounds. For example, a typical result produces a few large clusters, each containing very different compounds, and many singletons. There are sometimes clusters that contain compounds with a similarity of 0.2 on a scale of 0.0 to 1.0, with 1.0 the upper limit. Clearly, this is not clustering similar compounds together.

Much better results are found, albeit with datasets of 1000, by using statistically based agglomerative clustering methods. For 1000 compounds, the clustering takes approximately one day and would scale roughly as the square of the number of compounds. Since we typically would expect to investigate no more than 1/100 as many clusters as original compounds, divisive methods might have an advantage because in this approach, clustering starts with one huge cluster, divides clusters into tighter ones, and could stop once the target number of clusters was formed.

At this time, no method other than Jarvis-Patrick is known to the computational chemistry community that will group 10^5 or 10^6 objects in a time scale of less than a week (Willett et al., 1986; Willett, 1987; Whaley and Hodes, 1991). There are unpublished reports that the divisive Guenoche (1991) algorithm classifies 10^4 compounds overnight on a personal computer once the pairwise similarities have been calculated. However, it seems possible that there may be better ways to discover the groups of compounds in a dataset.

References

Baum, R.M., 1994, Combinatorial approaches provide fresh leads for medicinal chemistry, *Chemical and Engineering News* (February 7) 20-26.

Bron, C., and J. Kerbosch, 1973, Algorithm 457. Finding all cliques of an undirected graph. *Commun. ACM* 16:575.

Cramer III, R.D., D.E. Patterson, and J.D. Bunce, 1988, Comparative molecular field analysis (CoMFA). 1. Effect of shape on binding of steroids to carrier proteins, *J. Am. Chem. Soc.* 110:5959–5967.

Erickson, J.W., and S.W. Fesik, 1992, Macromolecular X-ray crystallography and NMR as tools for structure-based drug design, In *Annual Reports in Medicinal Chemistry,* Vol. 27, M.C. Vernuti, ed., Academic Press, New York, pp. 271–289.

Frank, I.E., and J.H. Friedman, 1993, A statistical view of some chemometrics regression tools, *Technometrics* 35:109–135.

Gallop, M.A., R.W. Barrett, W.J. Dower, S.P.A. Fodor, and E.M. Gordon, 1994, Applications of combinatorial technologies to drug discovery. 1. Background and peptide combinatorial libraries, *J. Medicinal Chem.* 37:1233–1251.

Gordon, E.M., R.W. Barrett, W.J. Dower, S.P.A. Fodor, and M.A. Gallop, 1994, Applications of combinatorial technologies to drug discovery. 2. Combinatorial organic synthesis, library screening strategies, and future directions, *J. Medicinal Chem.* 37:1387–1401.

Guenoche, A., P. Hansen, and B. Jaumard, 1991, Efficient algorithms for divisive hierarchical clustering with diameter criterion, *J. Classification* 8:5–30.

Hodes, L., 1989, Clustering a large number of compounds. 1. Establishing the method on an initial sample, *J. Chem. Inform. Comput. Sciences* 29:66–71.

Jarvis, R.A., and E.A. Patrick, 1973, Clustering using a similarity measure based on shared nearest neighbors, *IEEE Trans. Comput.* C-22:1025–1034.

Kubinyi, H., 1993, 3D QSAR in drug design, *Theory, Methods, and Applications,* ESCOM, Leiden, 759 pp.

Kuntz, I.D., 1992, Structure-based strategies for drug design and discovery, *Science,* 257:1078–1082.

Martin, Y.C., 1992, 3D database searching in drug design, *J. Medicinal Chem.* 35:2145–2154.

Martin, Y.C., K.-H. Kim, and C.T. Lin, in press, Comparative molecular field analysis: CoMFA, in *Linear Free Energy Relationships in Biology*, M. Charton, ed.

Rothstein, S.H., and M.A. Murcko, 1993, GroupBuild: A fragment-based method for de novo drug design, *J. Medicinal Chem.* 36:1700–1710.

Science, 1994, Research news: Drug discovery on the assembly line, 264:399–1401.

Weiniger, D., C. James, and J. Yang, 1994, *Daylight Chemical Information Systems*, Manual to Version 4.34, Daylight Chemical Information Systems, Irvine, Calif.

Whaley, R., and L. Hodes, 1991, Clustering a large number of compounds. 2. Using the connection machine, *J. Chem. Inform. Comput. Sciences* 31:345–347.

Willett, P., 1987, *Similarity and Clustering in Chemical Information Systems*, Research Studies Press, Letchworth.

Willett, P., V. Winterman, and D. Bawden, 1986, Implementation of nonhierarchic cluster analysis methods in chemical information systems, *J. Chem. Inform. Comput. Sciences* 26:109–118.

Statistical Analyses of Families of Structures

The diversity of chemical structures is one of the hallmarks of modern experimental chemistry. The problems of diversity and similarity are most prevalent in the study of biological molecules for which very different sequences—that is, fundamental structures—give rise to molecules that have very similar overall three-dimensional structures and often very similar functional properties. The Human Genome Project is devoted to characterizing the myriad of proteins encoded in humans, but a still larger universe of proteins exists in other living beings. Furthermore, it is easy to understand that the existing proteins are just a small subset of all possible random heteropolymers. The same type of combinatorial complexity exists for many other classes of molecules. In nature, we are familiar with the complexity of alkaloids or terpenes.

More and more molecular scientists are trying to understand how to use the information from a variety of different molecules to understand the structure and function of a given one. It is now becoming possible, by using combinatorial syntheses in the laboratory, to make 10 million variants of a single protein or 10,000 covalently connected frameworks such as those in a natural product. The most well known of these techniques is that employed to make catalytic antibodies, but many other approaches are possible. A variety of mathematical problems arise when one tries to make use of these resulting longitudinal data about molecular systems.

For naturally occurring biomolecules, one of the most important approaches is to understand the evolutionary relationships between macromolecules. This study of the evolutionary relationship between biomolecules has given rise to a variety of mathematical questions in probability theory and sequence analysis. Biological macromolecules can be related to each other by various similarity measures, and at least in simple models of molecular evolution, these similarity measures give rise to an ultrametric organization of the proteins. A good deal of work has gone into developing algorithms that take the known sequences and infer from these a parsimonious model of their biological descent.

Similar analyses based on the three-dimensional structure of molecules also present ongoing mathematical problems. At the moment, the use of evolutionary similarity to infer three-dimensional structure is a common and very important algorithm for people who have practical interests in the prediction of biomolecular structure. Use of the theory of spin glasses to characterize random heteropolymers has also allowed the phrasing of interesting questions such as the probability in a single experiment of obtaining a foldable protein molecule. This is a question in which the statistics of low-lying energy states on the surface and the statistics of sequences must be analyzed jointly and related to each other. Experiments of this type have recently been done and seem to agree in many respects with the results of theory, but there are many questions of physicochemical principle and of mathematical analysis for this theory.

An emerging technology is the use of multiple rounds of mutation, recombination, and selection to obtain interesting macromolecules or combinatorial covalent structures. Very little is understood as yet about the mathematical constraints on finding molecules in this way, but the mathematics of such artificial evolution approaches should be quite challenging. Understanding the navigational problems in a high-dimensional sequence space may also have great relevance to understanding natural evolution. Is it punctuated or is it gradual as many have claimed in the past? Artificial evolution approaches may obviate the need to completely understand and design biological macromolecules, but there will be a large number of interesting mathematical problems connected with the design of efficient artificial evolution experiments.

Quantum Monte Carlo Solution of the Schrödinger Equation

Many-body problems in physics are often treated by a Monte Carlo (MC) approach (e.g., Hammersley and Handscomb, 1964; Kalos and Whitlock, 1986). The Monte Carlo method is statistical and draws its name from the famous gambling casinos of Monaco because of the role of random numbers or coin tosses in the method.

Problems handled by Monte Carlo are generally of two types, probabilistic or deterministic, depending on whether they are connected with random processes. In the probabilistic case, the simple Monte Carlo approach is to observe the occurrence of random numbers, chosen in a way that they directly simulate the physical random processes of the original problem, and to infer the desired solution from the behavior of these random numbers. In the deterministic case, the power of the Monte Carlo approach is the capability of carrying out numerical calculations in cases where the equations that describe the essence of a problem and its underlying structure are not solvable by alternative means. The underlying structure or formal expression also describes some unrelated random process, and therefore the deterministic problem can be solved numèrically by a Monte Carlo simulation of the corresponding probabilistic problem.

The essential feature common to all Monte Carlo computations is that at some point one will need to substitute for a random variable a corresponding set of values with the statistical properties of the random variable. The values that are substituted are called random numbers. They are not really random, however, because if they were it would be impossible to repeat a particular run of a computer program. An absolute requirement in debugging a computer code is the ability to repeat a particular run of the program. If real random numbers were used, no calculation could be repeated exactly, and attempts to check for errors would be extremely difficult. It is essential that one be able to repeat a calculation when program changes are made or when the program is moved to a new computer.

For electronic computation it is desirable to calculate easily by a completely specified rule a sequence of numbers as required that will satisfy reasonable statistical tests for randomness for the Monte Carlo problem of interest. Such a sequence is called pseudorandom and clearly cannot pass every possible statistical test.

Most of the pseudorandom number generators now in use are special cases of the relation (Heermann, 1986; Kalos and Whitlock, 1986)

$$x_{n+1} \equiv a_0 x_n + a_1 x_{n-1} + \cdots + a_j x_{n-j} + b \qquad (\text{mod } P).$$

One initiates the generator by starting with a vector of $j + 1$ numbers $x_0, x_1, \ldots, x_j$. The generators are characterized by a period τ that in the best case cannot exceed P^{j+1}. The length of τ and the statistical properties of the pseudorandom sequences depend on the values of a_j, b, and P.

With the choice of $a_j = 0, j \geq 1$, and $b = 0$, one obtains the *multiplicative congruential generator*,

$$x_{n+1} \equiv \lambda \cdot x_n \qquad (\text{mod } P).$$

Recent work has shown that a set of parameters λ and P can be chosen with confidence to give desired statistical properties in many-dimensional spaces (Kalos and Whitlock, 1986). There are many other generators available; the interested reader should consult the references.

With parallel computers and supercomputers capable of very large calculations, very long pseudorandom sequences are necessary. In addition, there remains the desire to have reproducible runs. The question of independence of separate sequences to be used in parallel remains a research issue.

Quantum Monte Carlo (QMC), as used here, refers to a set of methods to solve the Schrödinger equation (exactly, in two of the three variants listed below) to within a statistical error by random walks in the many-dimensional space. These methods—variational MC, diffusion MC, and Green's-function MC—are based on the formal similarity between the Schrödinger equation in imaginary time and a multidimensional classical diffusion equation.

Variational Monte Carlo (VMC)

For $\Psi_T(\mathbf{R})$ a known approximate (trial) wavefunction, where $\mathbf{R}$ is the $3N$ set of coordinates of the N-particle system, VMC (Kalos and Whitlock, 1986; Hammond et al., 1994) uses the Metropolis algorithm (Hammersley and Handscomb, 1964; Heermann, 1986; Kalos and Whitlock, 1986; Hammond et al., 1994) to sample $|\Psi_T|^2$. Therefore, any expectation value with respect to this trial function can be computed including the variational energy of Ψ_T. To begin a calculation, an initial distribution of walkers is generated. In order to create a distribution of $|\Psi_T|^2$, these walkers take a series of Metropolis steps to equilibrate, followed by another series of Metropolis steps at which the local energy

$$E_{\text{local}} \equiv H\Psi_T(\boldsymbol{R})/\Psi_T(\boldsymbol{R})$$

is calculated for each walker; here H is the Hamiltonian. Averaging the local energy over the walkers at the sampling points yields the variational energy, which is an upper bound to the exact energy of the ground state. An alternative strategy is to minimize the variance of the local energies. A strength of the VMC method is the capability of treating trial functions that depend explicitly on interelectronic coordinates—there is no integral problem associated with the use of trial functions containing such coordinate dependence for many-electron systems.

Diffusion Monte Carlo (DMC)

If one multiplies the time-dependent Schrödinger equation in imaginary time by Ψ_T and rewrites it in terms of a new probability distribution $f(\boldsymbol{R},t) \equiv \int \Phi(\boldsymbol{R},t)\Psi_T(\boldsymbol{R})$, one obtains

$$\partial f/\partial t = \sum_i D_i \nabla_i(\nabla_i f - f\nabla_i \ln|\Psi_T|^2) - (E_L(\boldsymbol{R}) - E_T)f,$$

where $D_i \equiv \hbar^2/2m_i$, E_L is the local energy, and E_T (the trial energy) represents a constant shift in the zero of energy. At large simulation (imaginary) time, the function $\Phi(\boldsymbol{R},t)$ tends to the ground state wavefunction.

The algorithm (Hammond et al., 1994) is initiated with a distribution (ensemble) of several hundred walkers taken from $f(\boldsymbol{R},0) = |\Psi_T(\boldsymbol{R})|^2$, which is then evolved forward in time after a sequence of equilibration steps. The three terms on the right-hand side then correspond to diffusion with diffusion constant D_i, a drift term associated with the trial function, and a branching term that derives this designation from the DMC equation being, in the absence of the first two terms on the right-hand side,a first-order kinetic equation. Because f can, in general, assume both positive and negative values, which would preclude the interpretation of f as a probability for fermion systems, one alternative is to impose the nodes of the ground state wavefunction Ψ_T on Φ so that f is always positive. This is the fixed-node approximation, which can be shown to give an upper bound to the ground state energy.

After sufficiently long simulation time in which the steady-state solution of the DMC equation has been attained, the ground state energy is obtained as the average value of the local energy now averaged over the mixed distribution $\Phi(\boldsymbol{R})\Psi_T(\boldsymbol{R})$. This is an improvement over VMC and associated sampling from $|\Psi_T|^2$ because DMC contains information on the exact ground state Φ (in the fixed-node approximation). The ground state energy has the zero-variance property of QMC; that is, in the limit that Ψ_T approaches Φ, the variance of the MC estimate of the energy approaches zero.

Green's Function Monte Carlo (GFMC)

The integral equation form of the Schrödinger equation can also be modeled by a stochastic process, which leads immediately to the consideration of Green's functions. GFMC approaches (Kalos and Whitlock, 1986) have been investigated for the time-independent as well as the time-dependent Schrödinger equations. In practice, however, there are only convergent when the Fermi energy is close to the Bose energy and the trial function has sufficiently accurate nodes. New directions with GFMC that overcome these limitations still encounter limitations that restrict their application, at present, to first-row atoms. Nevertheless, GFMC remains an area for continued scrutiny.

Research Opportunities

The various forms of Monte Carlo for solving the Schrödinger equation (VMC, DMC, and GFMC) could each be improved by better sampling methods. Wavefunctions typically used for importance sampling often recover at best 80 to 95% of the correlation energy, the energy difference between the Hartree-Fock approximation and the nonrelativistic limit, for molecules consisting of first-row atoms. Beyond the first row, computer time dependence on atomic number, which has been estimated as z^{α} where $\alpha = 5.5$ to 6.5, is a major limiting factor that has led to the introduction of analytical functions to describe inner-shell electrons (i.e., pseudopotentials or effective core potentials). Another area in which improvements are eagerly sought is in methods that go beyond the fixed-node approximation. Currently, these methods are limited to atoms and molecules containing no more than 6 to 10 electrons.

References

Hammersley, J.M., and D.C. Handscomb, 1964, *Monte Carlo Methods*, Methuen, London.

Hammond, B.L, W.A. Lester, Jr., and P.J. Reynolds, 1994, *Monte Carlo Methods in Ab Initio Quantum Chemistry*, World Scientific, Singapore.

Heermann, D.C., 1986, *Computer Simulation Methods*, Springer-Verlag, Berlin.

Kalos, M.H., and P.A. Whitlock, 1986, *Monte Carlo Methods*, Volume 1: Basics, John Wiley & Sons, New York.

Nonadiabatic Phenomena[2]

The Born-Oppenheimer (BO) adiabatic approximation is the basis of the well-known separation of the many-body problem of electronic and nuclear motion into two separate many-body problems. It is

[2]This presentation follows Hirschfelder and Meath (1967).

also applied to separate other "fast" and "slow" subsystems including vibrational and rotational modes of molecules. The combination of these approximations leads to the double adiabatic approximation that has been applied in the study of solids as well as molecules.

For most applications, the approximation of separation of electronic and nuclear motion does not lead to appreciable error. For high precision, however, the BO energy has to be corrected for the coupling of electronic and nuclear motions. The coupling is important if, for example,

- two potential energy surfaces of the same symmetry cross in the BO approximation— correction terms prevent such crossings—or are close and more-or-less parallel over a moderate range of configuration space;
- the electronic state of a polyatomic molecule is degenerate for a symmetric arrangement of nuclei; coupling leads to Jahn-Teller and Renner-Teller effects;
- the nuclear velocities are large as in high-energy molecular collisions;
- the molecule is in a high rotational state and has nonzero electronic angular momentum.

There are two kinds of correction terms for the coupling of electronic and nuclear motions. The diagonal corrections shift the energy levels. The nondiagonal corrections produce and broaden the natural line width to the energy levels and cause transitions between quantum states. The energy corrected for the diagonal coupling terms is called the adiabatic energy and gives the best possible energy curves and surfaces.

A number of problems arise in the inclusion of nondiagonal corrections. They occur because of divergences in coupling matrix elements in regions where potential energy surfaces approach very closely or because the BO electronic basis functions may not be appropriate for the region of close approach.

BO deviations arise for two reasons. First, coupling terms appear in the kinetic energy when the coordinates are transformed from the laboratory-fixed axes to the body-fixed (molecular) axes. Second, the Breit-Pauli relativistic corrections to the electrostatic Hamiltonian lead to spin-spin, spin-orbit, and other magnetic coupling terms. The present discussion neglects the latter and focuses on the former.

The Hamiltonian in the laboratory-fixed frame may be written

$$H' = -\frac{1}{2}[\sum_i \nabla'^2_i + \sum_\alpha \nabla'^2_\alpha / m_\alpha] + U \,, \tag{21}$$

where primes denote the laboratory frame, m_α is the mass of the α-th nucleus in units of electron mass, and U is the potential energy given by the Coulomb interactions (nuclear, electronic-nuclear, and electronic) of all the particles of the system. Since U is a function only of relative distances between the particles, one separates out the center-of-mass motion and is left with $3(n+N) - 3$ relative coordinates for an n-electron and N-nucleus molecular system. The choice of optimum coordinates for polyatomic molecules is not straightforward, and so the focus here, for simplicity, is on the diatomic molecule case. Then the Hamiltonian takes the form

$$H = H_e - \nabla^2_R/(2\mu) - \frac{1}{2}[\sum_i \nabla^2_i I + 2\sum_{i<j} \nabla_i \cdot \nabla_j] \,, \tag{22}$$

where $\mu = m_a \cdot m_b/(m_a + m_b)$ and

$$H_e = -\frac{1}{2}\sum_i \nabla_i^2 - \sum_i \left[\frac{Z_a}{r_{ai}} + \frac{Z_b}{r_{bi}}\right] + \sum_{i<j} \frac{1}{r_{ij}} + \frac{Z_a Z_b}{R}. \tag{23}$$

Here Z_a and Z_b are the nuclear charges, r_{ai} is the distance separating nucleus a from electron i, and R is the distance between nuclei a and b.

If one assumes that the electronic Schrödinger equation,

$$H_e \psi_k(\boldsymbol{r},\boldsymbol{R}) = E_k(\boldsymbol{R})\psi_k(\boldsymbol{r},\boldsymbol{R}), \tag{24}$$

can be solved exactly for the complete set of eigenfunctions $\psi_k(\boldsymbol{r},\mathbf{R})$ and eigenvalues $E_k(\boldsymbol{R})$, then the total Schrödinger equation for nuclear and electronic motions,

$$H\Psi(\boldsymbol{r},\boldsymbol{R}) = E\Psi(\boldsymbol{r},\boldsymbol{R}), \tag{25}$$

can be solved by expanding Ψ as follows,

$$\Psi(\boldsymbol{r},\boldsymbol{R}) = \sum_k \phi_k(\boldsymbol{R})\psi_k(\boldsymbol{r},\boldsymbol{R}). \tag{26}$$

Here $\boldsymbol{r}$ represents all the coordinates of the electrons and k is the set of electronic quantum numbers. This leads to a set of equations for the functions $\phi_k(\boldsymbol{R})$ that determine the nuclear motion of the system,

$$\begin{aligned}[-\nabla_R^2/(2\mu) + E_\ell(\boldsymbol{R}) + E_{\ell\ell}'(\boldsymbol{R}) + E_{\ell\ell}''(\boldsymbol{R})\cdot\nabla_R - E]\,\phi_k(\boldsymbol{R}) \\ = -\sum_{k\neq\ell}[E_{\ell k}'(\boldsymbol{R}) + E_{\ell k}''(\boldsymbol{R})\cdot\nabla_R]\,\phi_k(\boldsymbol{R}).\end{aligned} \tag{27}$$

Here

$$E_{\ell k}'(\boldsymbol{R}) = E_{\ell k}'(\nabla_R^2) + E_{\ell k}'(\nabla_i\cdot\nabla_j) + E_{\ell k}'(\nabla_i^2),$$

where

$$E_{\ell k}'(\nabla_R^2) = -1/(2\mu)\int \psi_\ell^*(\boldsymbol{r},\boldsymbol{R})\,\nabla_R^2\,\psi_k(\boldsymbol{r},\boldsymbol{R})\,d\boldsymbol{r},$$

$$E_{\ell k}'(\nabla_i^2) = -1/[2(m_a+m_b)]\sum_i \int \psi_\ell^*(\boldsymbol{r},\boldsymbol{R})\,\nabla_i^2\,\psi_k(\boldsymbol{r},\boldsymbol{R})\,d\boldsymbol{r},$$

$$E_{\ell k}'(\nabla_i\cdot\nabla_j) = -1/(m_a+m_b)\sum_{i<j}\int \psi_\ell^*(\boldsymbol{r},\boldsymbol{R})\,\nabla_i\cdot\nabla_j\psi_k(\boldsymbol{r},\boldsymbol{R})\,d\boldsymbol{r},$$

and

$$E_{\ell k}''(\boldsymbol{R}) = -1/\mu\int \psi_\ell^*(\boldsymbol{r},\boldsymbol{R})\,\nabla_R\psi_k(\boldsymbol{r},\boldsymbol{R})\,d\boldsymbol{r}. \tag{28}$$

An asterisk denotes complex conjugate. The quantities defined by Equation (28) give rise to velocity-dependent forces on the nuclei. For real ψ_ℓ, however, the diagonal term vanishes.

In practice, Equation (27) is extremely difficult to solve and various approximations to it are introduced. The BO approximation corresponds to neglecting all of the coupling terms. Equation (27) then becomes a Schrödinger equation for nuclear motion,

$$[-1/(2\mu)\nabla_R^2 + E_\ell(\boldsymbol{R}) - E]\phi_\ell(\boldsymbol{R}) = 0 .$$

In this approximation, the $E_\ell(\mathbf{R})$ determined from Equation (24) become the potential energy for the nuclear motion.

The adiabatic approximation corresponds to neglecting all nondiagonal terms in Equation (27), which results in a Schrödinger-type equation for nuclear motion,

$$[-1/(2\mu)\nabla_R^2 + V_\ell(\boldsymbol{R}) - E]\phi_\ell(\boldsymbol{R}) = 0,$$

where the potential for nuclear motion is

$$V_\ell(\boldsymbol{R}) = E_\ell(\boldsymbol{R}) + E_{\ell\ell}'(\boldsymbol{R}) .$$

The diagonal elements $E_{\ell\ell}'(\boldsymbol{R})$ are effectively a correction to the potential energy due to the coupling between the electronic and nuclear motions.

The adiabatic approximation gives the best potential energy function. As defined here the adiabatic approximation to the energy is an upper bound to the true energy since it can be expressed as the expectation value to the correct Hamiltonian for the molecule evaluated with an approximate wave function.

The nonadiabatic approximation corresponds to consideration of the nondiagonal as well as the diagonal elements $E_{\ell k}'(\boldsymbol{R})$. This is extremely difficult to carry out but has been accomplished for H_2 by using variational basis set (Kolos and Wolniewicz, 1960) and quantum Monte Carlo approaches (Traynor et al., 1991; see also Ceperly and Alder, 1987).

For processes occurring in excited states, it appears clear that for molecules more complicated than H_2 alternative approaches must be found. This is an area demanding fundamental improvements if breakthroughs are to be achieved.

References

Ceperley, D.M., and B.J. Alder, 1987, *Phys. Rev. B* 36:2092.

Hirschfelder, J.O., and W.J. Meath, 1967, *Adv. Chem. Phys.* 12:1.

Kolos, W., and L. Wolniewicz, 1960, *Rev. Mod. Phys.* 35:1323.

Traynor, C.A., J.B. Anderson, and B.M. Boghosian, 1991, A quantum Monte-Carlo calculation of the ground-state energy of the hydrogen molecule, *J. Chem. Phys.* 94:3657–3664.

Evaluation of Integrals with Highly Oscillatory Integrands: Quantum Dynamics with Path Integrals

The solutions of many quantum dynamics problems can be formulated in terms of path integrals. Indeed, there is a formulation of quantum mechanics (less familiar than the Schrödinger wave function formulation, the Heisenberg matrix formulation, or the Dirac Hilbert space formulation) based entirely on the use of path integrals (Feynman and Hibbs, 1965). Path integrals are closely related to some of the mathematical approaches used to describe classical Brownian motion. A prototypical mathematical problem and a class of methods for solving it are described first; then a more general discussion of the problem is presented.

Prototypical Problem

This discussion of the prototypical problem follows closely the discussion of Doll et al. (1988). Consider the integral $I(t) = \int d\mathbf{x}\, \rho(\boldsymbol{x})\, e^{itf(x)}$ where t is a real parameter corresponding to the time, $\boldsymbol{x}$ is the set of coordinates for a many-dimensional space, and $\rho(\boldsymbol{x})$ and $f(\boldsymbol{x})$ are functions in that space. The function $\rho(\boldsymbol{x})$ is positive semidefinite and can without loss of generality be taken as normalized to unity. Thus, it can be regarded as a probability density in the space. The function $f(x)$ is real (for this prototype problem).

For problems of interest, the dimensionality of the space can be very large (from 1 to a hundred to thousands), and the complexity of the function precludes analytic integration. For any particular $\boldsymbol{x}$, $\rho(\boldsymbol{x})$ and $f(\boldsymbol{x})$ can be evaluated numerically, and typically there is much more computation required for the evaluation of ρ than f.

The range of the $\boldsymbol{x}$ integration may be bounded, but more often it is unbounded. For typical problems of interest, however, integrals of the form $\int dx\, \rho(\mathbf{x})\, (f(\mathbf{x}))^n$ exist for all positive integers n.

The goal is usually to evaluate the integral for a specific value of t (or for a set of values of t) by making use of calculated values of ρ and f at appropriately chosen points and to estimate the accuracy of the answer. In some cases, the desired quantity is the area under the function

$$\int_{-\infty}^{\infty} dt I(t),$$

or perhaps the Fourier transform

$$\int_{-\infty}^{\infty} dt I(t) e^{i\omega t}.$$

Techniques for evaluation of such integrals of I would be extremely worthwhile, but this discussion focuses on the evaluation of $I(t)$ for specific values of t.

Discussion of the Problem

A standard way of attacking this problem would be a Monte Carlo integration scheme in which points $\boldsymbol{x}$ are generated with a probability distribution of $\rho(x)$ and f is evaluated at these points. Then

$$I(t) \approx \frac{1}{N}\sum_{n=1}^{N} e^{itf(x_n)},$$

where N points are generated and $\boldsymbol{x}_n$ is the nth point. For small values of t, this can be a practical procedure. For large values of t, however, this procedure is extremely inefficient because of the large amount of cancellation among the various terms.

If t is large and f is not a constant, it is clear that the integrand is highly oscillatory; thus, small regions of space where f is varying will tend to contribute small net amounts to the integrand for large t. Nevertheless, individual points in such a region will contribute to the sum an exponential whose magnitude is unity. Cancellation among the various terms will give an accurate and small answer only when each such small region is sampled many times in the Monte Carlo evaluation.

The regions of the space in which $df(x)/dx$ is zero or small will make the major contributions to the integral for large t. Thus, a scheme that concentrates the Monte Carlo sampling at and near such regions would be desirable. Such schemes are usually called "stationary-phase" methods, since at such points the phase of the exponential is stationary.

Stationary-Phase Monte Carlo Methods

A variety of sampling methods have been developed that focus on the stationary-phase points (Doll, 1984; Filinov, 1986; Doll et al., 1988). Here the approach of Doll et al. (1988), which is similar to the others, is outlined.

The integral of interest can be rewritten identically as

$$I(t) = \int dx\, \rho(x) D(x,t)\, e^{itf(x)}, \tag{29}$$

where

$$D(x,t) = \int dy\, P(y) \frac{\rho(x-y)}{\rho(x)} e^{it[f(x-y)-f(x)]},$$

and $P(y)$ is an arbitrary normalized probability distribution. (This result is exact if the range of integration is over all positive and negative values of the integration variables.) The function $P(y)$ is typically chosen to be a Gaussian distribution centered at $y = 0$. $D(x,t)$ is called a "damping function," because for such a typical choice of $P(y)$, $D(x,t)$ is largest where f is stationary and hence it damps the integrand in Equation (29) away from the stationary points.

An evaluation of $D(x,t)$ poses the same problems as evaluation of $I(t)$. However, for specific choices of P, various approximations for D can be constructed and used as the basis for a more efficient sampling method for the evaluation of I. For example, if $P(y)$ is a narrow Gaussian function with a maximum at $y = 0$, then for large t an approximation can be constructed by replacing $\rho(x - y)/\rho(x)$ by unity and performing a Taylor expansion of the exponent. One then obtains an approximation for D called the "first-order gradient approximation":

$$D(x,t) \approx D_0(x,t) = \exp[-(\epsilon t f'(x))^2/2]\,.$$

Then

$$I(t) \approx \int dx\, \rho(x) D_0(x,t)\, e^{itf(x)}\,.$$

This integral, which is the major contribution to the correct answer, is generally more amenable to evaluation by sampling, in this case by using $\rho(x)D_0(x,t)$ as the probability distribution. A complete method would require the development of procedures to estimate the correction terms that must be added on the right to make this an equality (see Doll et al., 1988, for additional details). The essential idea of these procedures is that use of a precise approximate formula for D automatically includes much of the cancellation responsible for decreasing the value of I.

The stationary-phase Monte Carlo methods provide some useful ideas for formulating tractable sampling methods for solving some problems of interest. It would be worthwhile to develop better versions of these methods both for $I(t)$ and for its Fourier transform.

Alternative Approaches to the Prototype Problem

In many cases of interest, the function $I(t)$ is known to be an analytic function of t at the origin. This knowledge may be of use in developing approximate evaluation techniques based on analytic continuation. Most analytic continuation methods are based on a more detailed formulation of the problem of interest that is tied closely to the physical nature of the problem. But even for this simple formulation of the prototype problem, one might ask whether the fact that $I(t)$ is an analytic function of (complex) t for some region that includes the origin suggests any alternate ways of calculating $I(t)$ accurately.

For example, one might consider the use of Padé approximants (i.e., ratios of polynomials; Baker and Gammel, 1970), which are often used for approximate analytical continuation of analytic functions, especially meromorphic functions. Expanding the exponential in $I(t)$ gives

$$I(t) = \sum_{n=0}^{\infty} (it)^n \alpha_n / n!,$$

where

$$\alpha_n = \int dx\, \rho(x)(f(x))^n .$$

The first few Taylor series coefficients can then be estimated by sampling on $\rho(x)$ in the usual way, but the difficulty of making accurate estimates grows as n increases. These coefficients might then be used to fit $I(t)$ with a Padé approximant.

This approach raises a number of mathematical questions.

- Are there any sampling techniques that would be especially effective in evaluating the coefficients for large n?
- Does the statistical error in the α_n make it inappropriate to use Padé approximants to fit the function? Are there better alternatives?
- If the Padé approximant method is valid, how is the uncertainty in $I(t)$ related to the statistical uncertainty of the individual α_n?
- Does the fact that the integral expression for $I(t)$ is dominated by stationary points in the $\mathbf{x}$ space give any insights into how to evaluate α_n or how to construct appropriate Padé approximants?

Other Formulations and Solutions of the Basic Problem

The nature and dimensionality of the $\mathbf{x}$ space and the specific form of the $\rho(\mathbf{x})$ function are determined by the nature of the quantum mechanical problem to be solved. In general, however, the coordinates of $\mathbf{x}$ are Cartesian (or other) coordinates for a mechanical system. Similarly, the form of the function $f(\mathbf{x})$ is determined by the nature of the problem. Various analytic continuation techniques have been applied to attack specific special cases of the prototype problem. The question then arises as to whether these are isolated solutions of specific problems or special cases of an underlying, not yet discovered, general method for solving problems of this type. This is not the place to delve into the details of specific problems; instead, some of the relevant mathematical principles involved are highlighted. For additional discussion of related problems, the reader is referred to Doll (1984), Makri (1991), and Wolynes (1987).

Analytic Continuation in Time. In some problems, the goal is to evaluate a function of time t of the form

$$h(t) = \frac{\int dx\, e^{g(it,x)} f(x)}{\int dx\, e^{g(it,x)}} . \tag{30}$$

More generally, there is a function of a complex variable z of the form

$$H(z) = \frac{\int dx\, e^{g(z,x)} f(x)}{\int dx\, e^{g(z,x)}} \tag{31}$$

that is analytic in a region including the positive real axis and part of the positive imaginary axis, and for which $h(t) = \lim_{\epsilon \to 0+} H(\epsilon + it)$. The function $g(z,x)$ is an analytic function of all its arguments, and $f(x)$ is real (for real x). In many cases, the formulation in Equation (31) is required to make the problem well defined, because $g(it,x)$ is purely imaginary and the integrals in Equation (30) are not absolutely convergent (for an infinite range of integration).

The calculation of $H(z)$ for real z presents a tractable sampling problem because $g(z,x)$ is real for real z. Then, sampling a set of points distributed with a probability density proportional to $g(z,x)$ and calculating f at those points allow H to be calculated for real z. In fact, as long as z has a positive real part, the problem of evaluating H can be converted to a well-defined sampling problem, because $\Re g(z,x) \to -\infty$ for large positive and negative values of the integration variables. Then $\Re g(z,x)$ can be used as a probability distribution, and the right side of Equation (31) can be expressed as a ratio of two averages over this distribution:

$$\begin{aligned} H(z) &= \frac{\int dx\, e^{\Re g(z,x)} e^{i\Im g(z,x)} f(x)}{\int dx\, e^{\Re g(z,x)} e^{i\Im g(z,x)}} \\ &= \frac{\int dx\, e^{\Re g(z,x)} e^{i\Im g(z,x)} f(x) \,/ \int dx\, e^{\Re g(z,x)}}{\int dx\, e^{\Re g(z,x)} e^{i\Im g(z,x)} \,/ \int dx\, e^{\Re g(z,x)}} . \end{aligned}$$

For complex values of z, however, this sampling problem may become intractable because the quantities to be averaged, $\exp\,[i\Im g(z,x)]\, f(x)$ and $\exp[i\Im g(z,x)]$, are complex and oscillatory. This can pose the same difficulty as that discussed for the prototypical problem above. This difficulty becomes acute in the limit that z becomes purely imaginary, because the "probability distribution function" approaches unity and is not normalizable.

A variety of analytic continuation methods have been used to solve problems of this type. One strategy is to perform tractable sampling calculations to obtain $H(z)$ for real z and then use Padé approximant techniques to estimate H as z approaches the imaginary axis (Miller et al., 1983; Thirumalai and Berne, 1983; Jacquet and Miller, 1985; Yamashita and Miller, 1985; Makri and Miller, 1989). These techniques motivate the same sorts of mathematical questions as those listed above for a different approach to analytic continuation.

Deformation of the Integration Contours. Another technique has been used in certain situations in which the integrand is an analytic function of all the variables in $\boldsymbol{x}$. The integration is from $-\infty$ to ∞ along the real line for each of these variables. Deformation of these contours of integration and evaluation of the resulting integrals by sampling methods can be an effective technique if the deformed contours come near or pass through stationary-phase points of the integrand (Chang and Miller, 1987; Mak and Chandler, 1991). In one situation in which this was used (Mak and Chandler, 1991), the contour was described by a set of parameters that were varied during the calculation to maximize some measure of the success of the sampling process being used to calculate the integral.

Most General Formulation. Perhaps the most general formulation of these problems is the following one. Given a set of complex variables $z_1, \ldots, z_n$ and a function of these variables $g(z_1, \ldots, z_n)$ that is an analytic function of all these variables for a domain that includes the real axis for each of them, calculate the following complex function of m complex variables:

$$H(z_1,\ldots,z_m) = \int_{-\infty}^{\infty} dz_{m+1} \int_{-\infty}^{\infty} dz_{m+2} \cdots \int_{-\infty}^{\infty} dz_n g(z_1,\ldots,z_n) .$$

The method should work for highly oscillatory functions and should take into account the fact that each evaluation of g and the location of stationary-phase points are expensive. Here $z_1, \ldots, z_m$ are analogous to the t in previous examples, and $z_{m+1}, \ldots, z_n$ are the integration variables.

References

Baker, Jr., G.A., and J.L. Gammel, 1970, *The Padé Approximant in Theoretical Physics,* Academic Press, New York.

Chang, J., and W.H. Miller, 1987, Monte-Carlo path integration in real-time via complex coordinates, *J. Chem. Phys.* 87:1648.

Doll, J.D., 1984, Monte Carlo Fourier path integral methods in chemical dynamics, *J. Chem. Phys.* 81:3536.

Doll, J.D., Thomas L. Beck, and David L. Freeman, 1988, Quantum Monte-Carlo dynamics—The stationary phase Monte-Carlo path integral calculation of finite temperature time correlation-functions, *J. Chem. Phys.* 89:5753–5763.

Feynman, R.P., and A.R. Hibbs, 1965, *Quantum Mechanics and Path Integrals*, McGraw-Hill, New York.

Filinov, V.S., 1986, Calculation of the Feynman integrals by means of the Monte-Carlo method, *Nucl.Phys. B* 271:717–725.

Jacquet, R., and W.H. Miller, 1985, *J. Phys. Chem.* 89:2139.

Mak, C.H., and D. Chandler, 1991, Coherent-incoherent transition and relaxation in condensed-phase tunneling systems, *Phys. Rev.* 44:2352–2369.

Makri, N., 1991, Feynman path integration in quantum dynamics, *Computer Physics Communications* 63:389.

Makri, N., and W.H. Miller, 1987, Monte-Carlo integration with oscillatory integrands: implications for Feynman path integration in real time, *Chem. Phys. Lett.* 139:10–14.

Makri, N., and W.H. Miller, 1989, Exponential power series expansion for the quantum time evolution operator, *J. Chem. Phys.* 90:904–911.

Miller, W.H., S.D. Schwartz, and J.D. Tromp, 1983, Quantum mechanical rate constants for bimolecular reactions, *J. Chem. Phys.* 79:4889–90.

Thirumalai, D., and B.J. Berne, 1983, On the calculation of time correlation functions in quantum systems: path integral techniques, *J. Chem. Phys.* 79:5029–33.

Wolynes, P.G., 1987, Imaginary time path integral Monte Carlo route to rate coefficients for nonadiabatic barrier crossing, *J. Chem. Phys.* 87:6559.

Yamashita, K., and W.H. Miller, 1985, "Direct" calculation of quantum mechanical rate constants via path integral methods: Application to the reaction path Hamiltonian, with numerical test for the H + H_2 reaction in 3D, *J. Chem. Phys.* 82:5475–5484.

Fast Algebraic Transformation Methods

Numerous problems require rapid evaluation of a product of a matrix and a vector,

$$y_i = \sum_{j=1}^{n} a_{ij} x_j \qquad \forall i = 1, \ldots, n .$$

This arises in the transformation from one basis in a linear space to another. If computed as written, this transformation computes the n values of the transform in an amount of work proportional to n^2. However, there are both exact and approximate methods that can compute such a transform much more efficiently in a variety of cases.

One of the most famous is the Fourier transform, which is frequently used to transform from physical space to frequency space. The fast Fourier transform (FFT) is a well-known algorithm for computing the n values of the transform in an amount of work proportional to $n \log_2 n$. The FFT is an *exact* method, not an approximation: in exact arithmetic, it computes the sum exactly by a reordering of the summands. Although this algorithm is remarkably successful, recent advances have been made by further exploiting its algebraic structure.

Efficient implementation of large multidimensional FFTs on modern computers is memory bound. This can be seen from the following timings for a Sun Sparcstation 10 performing an FFT on a two-dimensional image of size 2K by 2K words using the standard FFT algorithm. When carried out on a machine with 64 megabytes of random access memory (RAM), the algorithm executes in 66 seconds. However, on a machine with only 32 megabytes of RAM, the algorithm takes more than two hours, due to the fact that the 2K by 2K dataset cannot fit in RAM and data must be moved in and out of memory during the calculation. Recently, J.R. Johnson and R.W. Johnson (private communication, June 1994) reported that a truly multidimensional version of the FFT algorithm has been developed that accomplishes the same computation in only 75 seconds on a 32-megabyte machine and 47 seconds on a 64-megabyte machine.

A recent example of approximate methods is the fast multipole expansion technique, which takes advantage of special properties of the matrix (a_{ij}). This technique has been successfully applied to evaluate Coulombic potentials in molecular dynamics simulations (Ding et al., 1992). This idea has been extended (Draghicescu, 1994) to a broad class of matrices using a general approximation procedure. These more general techniques would apply directly to the non-Coulombic potentials arising in periodic problems through the use of Ewald sums.

A closely related algebraic operation occurs in the evaluation of multilinear forms such as those that appear in density functional methods for calculating ground electronic states in quantum electronic structures. A bilinear form $B((z_i),(x_i))$ can always be written in terms of a matrix:

$$B((z_i),(x_i)) = \sum_{i,j=1}^{n} z_i a_{ij} x_j .$$

As written, the bilinear form would require $O(n^2)$ operations to evaluate. It can be computed instead by using the intermediate vector (y_i) defined above, calculation of which requires $O(n)$ operations, as

$$B((z_i),(x_i)) = \sum_{i=1}^{n} z_i y_i .$$

This way of evaluating a bilinear form can greatly reduce overall computation time, depending on how efficiently y_i can be computed.

In density functional methods for calculating ground electronic states in quantum electronic structures, it is desired to approximate multilinear forms including integrals of the form

$$\iint \frac{\rho(\boldsymbol{x})\,\rho(\boldsymbol{x}')}{|\boldsymbol{x}-\boldsymbol{x}'|}\, d\boldsymbol{x}\, d\boldsymbol{x}',$$

where

$$\rho(\boldsymbol{x}) = \sum_{j=1}^{n} C_{ij}\phi_i(\boldsymbol{x})\phi_j^*(\boldsymbol{x}) \quad \text{and} \quad \phi_i(\boldsymbol{x}) = \boldsymbol{x}^{\mu_i}\exp[-\gamma_i|\boldsymbol{x}-\boldsymbol{x}_i|]^2 .$$

(Here, μ_i denotes a multi-index, say (p_i, q_i, r_i), so that $\boldsymbol{x}^{\mu_i} = x_1^{p_i}\cdot x_2^{q_i}\cdot x_3^{r_i}$

In particular, such a representation arises via an expansion

$$\chi^\alpha(\boldsymbol{x}) = \sum_{i=1}^{n} c_i^\alpha \phi_i(\boldsymbol{x}) \qquad \text{and} \qquad \rho(\boldsymbol{x}) = \sum_{\alpha=1}^{m} |\chi^\alpha(\boldsymbol{x})|^2 .$$

Hence,

$$C_{ij} = \sum_{\alpha=1}^{m} (c_i^\alpha)^* c_j^\alpha .$$

The evaluation of ρ at a single point provides an example of the alternatives for evaluating a bilinear form. Using the representation involving the matrix (C_{ij}) requires n^2 operations, whereas the expression involving the χ's requires $m \cdot n$ operations. If $m \ll n$, the latter approach will be more efficient.

The above integral can be viewed as a quadrilinear form involving the coefficients c_i^α. Multilinear forms can be evaluated in a variety of ways. It is tempting to represent them in terms of a precomputed tensor (a matrix for a bilinear form). Recently, Bagheri et al. (1994) have observed that it can be more efficient, in both time and memory, not to precompute expressions.

The original integral can be written

$$\iint \sum_{\alpha=1}^{m} \left|\sum_{i=1}^{n} c_i^{\alpha} \phi_i(x)\right|^2 \cdot \sum_{\beta=1}^{m} \left|\sum_{j=1}^{n} c_j^{\beta} \phi_j x'\right|^2 \frac{dx\,dx'}{|x-x'|}.$$

By using a suitable quadrature rule, this could be approximated via

$$\sum_{\xi} \sum_{\xi'} \omega_{\xi} \omega_{\xi'} \sum_{\alpha=1}^{m} \left|\sum_{i=1}^{n} c_i^{\alpha} \phi_i(\xi)\right|^2 \sum_{\beta=1}^{m} \left|\sum_{j=1}^{n} C_j^{\beta} \phi_j(\xi')\right|^2.$$

Let $Q = q^2$ denote the total number of quadrature points; the quadrature approximation can be calculated in an amount of work proportional to Qnm.

On the other hand, the original integral can be expressed as

$$\iint \sum_{\alpha=1}^{m} \sum_{i,i'=1}^{n} (c_i^{\alpha})^* c_{i'}^{\alpha} \phi_i(x)^* \phi_{i'}(x) \sum_{\beta=1}^{m} \sum_{j,j'=1}^{n} (c_j^{\beta})^* c_{j'}^{\beta} \phi_j(x)^* \phi_{j'}(x) \frac{dx\,dx'}{|x-x'|}$$

$$= \sum_{i,i',j,j'=1}^{n} C_{i,i'} C_{j,j'} K_{i,i',j,j'},$$

where computation of the intermediate quantities

$$C_{i,i'} = \sum_{\alpha=1}^{m} (c_i^{\alpha})^* c_{i'}^{\alpha}$$

has been introduced, which adds only $O(n^2m)$ work. The summation over i, i', j, j' requires an amount of work and storage proportional to n^4 to evaluate. It would be more efficient to use the former technique if the quadrature could be done sufficiently accurately with $Q \ll n^3/m$ points.

The integral form involving ρ can be written as

$$\iint \frac{\rho(x)\rho(x')}{|x-x'|} dx\,dx' = \int \rho(x) K \rho(x) dx,$$

where K is an operator defined by

$$K\rho(x) \equiv \int \frac{\rho(x')}{|x-x'|} dx'.$$

If one thinks of an integral simply as a very large summation, the analogy between the integral form and the original bilinear form is apparent. With a rapid way to evaluate K, there might be a faster way to evaluate the bilinear form. (Alternative ways to evaluate ρ, based on similar considerations, have already been discussed.)

Observe that K is simply the solution operator for Poisson's equation

$$-\Delta K f = f.$$

With this in mind, there are numerous ways to evaluate the action of K approximately, using discrete methods for approximating the solution of Poisson's equation. For example, the multigrid method can be used to do this in a very efficient way. The analogy with quadrature suggests using a grid with q points and solving Poisson's equation with a multigrid method in $O(q)$ work. Then the complete evaluation could be done in $O(qmn)$ work. This potentially would be faster than direct evaluation of the integrals.

References

Bagheri, B., L.R. Scott, and S. Zhang, 1994, Implementing and using high-order finite element methods, *Finite Elements in Analysis and Design* 16:175–189.

Ding, Hong-Qiang, Naoki Karasawa, and William A. Goddard III, 1992, Atomic level simulations on a million particles: The cell multipole method for Coulomb and London nonbond interactions, *J. Chem. Phys.* 97:4309–4315.

Draghicescu, C.I., 1994, An efficient implementation of particle methods for the incompressible Euler equations, *SIAM J. Numer. Anal.* 31:1090–1108.

5
CULTURAL ISSUES AND BARRIERS TO INTERDISCIPLINARY WORK

The disparate natures of the fields of mathematics and chemistry have led to quite different training frameworks, which in turn continuously influence and are influenced by the research styles and practices that are most common in the two communities. Both fields are highly heterogeneous, containing many distinct research subfields, each with its own style, technical language, and point of view. While sweeping generalities are consequently inaccurate, broad stylistic differences can be identified in the ways in which mathematicians and chemists think, approach problems, and interact with their own and other communities. This chapter attempts to identify specific impediments to interactions between mathematics and chemistry, beyond the general factors affecting interdisciplinary research (see National Research Council and Institute of Medicine, 1990, for example).

Motivation and Connections

Many of the problems that interest and drive computational/theoretical chemists are practical (chemical syntheses, understanding of structure-function relationships in macromolecules, drug design, and so forth). To investigate these problems, chemists must build theories and perform numerical simulations to aid in understanding phenomena that can potentially be verified by experiment or yield new (predictive) data. The importance of mathematics in chemistry is revealed by the success of chemists in developing effective algorithms not only for specialized problems in chemistry, but also for generic mathematical problems—for example, evaluating integrals, calculating matrix eigenvalues, storing and compressing data, finding multidimensional optima and stationary points, and generating numerical solutions to the one-dimensional Schrödinger equation.

The key role of mathematics in computational chemistry highlights an anomaly: although theoretical chemists understand sophisticated mathematics and make heavy use of the mathematical literature, they have typically not involved mathematicians directly in either the development of models or algorithms or the derivation of formal properties of equations and solutions. In fact, theoretical chemists have become accustomed to self-reliance in mathematics. To date, this system has worked because most leading computational chemists have significantly more training in mathematics than the minimum recommended today for an American chemist (see Box 5.1).

By contrast to the situation in chemistry, the motivation for studying problems in much of pure mathematics does not depend on any connection to an application. Mathematical problems are often studied because of their inherent beauty, richness, and depth, without considering utility or relevance in any time frame. Even when addressing real-world applications, mathematicians tend to view problems as generic rather than specific.

In particular, "real" problems, whether from the biological or physical sciences, criminology, or cryptography, are almost always posed incompletely in a strictly mathematical sense. The transformation of a real-world problem into a tractable mathematical form involves increasing levels of abstraction and assumption; the physical terms are defined in a mathematical framework, and various components of the problem may be removed or idealized to build a model that emphasizes what are believed to be the most crucial features.

At its best—which is usually when the model building is done collaboratively, drawing on the particular insights and strengths of both mathematical and chemical scientists—this diagnostic approach is extremely valuable, leading to both theoretical and numerical understanding of the model, which can then return both quantitative information and conceptual understanding about the original physical problem. Indeed, carefully designed models can suggest and explain properties that are counterintuitive or unexpected to the problem posers themselves. An example is found in macroscopic models of supercoiled DNA, in which higher buckling catastrophes, as extensions to elasticity theory, were recently found. Furthermore, analysis of mathematical models can indicate directions for additional physical and numerical experimentation, as well as for extensions to the model. A potentially negative effect of mathematical abstraction, however, is that the theoretical formulation may lose its relevance to the original application that motivated it.

Effects of Disciplinary Boundaries

Beyond the "cultural" differences just described, attempts to build collaborations between mathematicians and chemists encounter boundaries imposed within most universities as well as within other structures in which disciplinary divisions are strong. These borders are especially difficult to cross early in a scientist's career, ironically when the potential for interdisciplinary work may be greatest because of the appeal of "new frontiers." Institutional practices influence the style of work that is valued (and hence often pursued) in each discipline, as well as the level of regular interaction and communication. Since many computational chemists and mathematicians are faculty members at research universities, both fields are affected by the value system of academia, in which recognition (promotion and tenure) requires a record of individual accomplishment judged as outstanding by one's peers. During the process of creating a portfolio of publications, any perceived dilution of a faculty member's personal contribution through collaborations may be seen as undesirable. This discouragement of collaborative work early in one's career applies to both mathematics and chemistry, although chemists have a strong countervailing tradition of working in groups.

For mathematicians, the potential career damage of collaboration rises when it involves work in a field seen as peripheral to mathematics. In some instances, interdisciplinary work may be regarded by one's mathematical colleagues as "not real mathematics" or as less valuable than traditional mathematics. Most academic mathematicians would agree that it is difficult to obtain accurate and convincing evaluations of "interdisciplinary" work (meaning work that involves significant contributions from other sciences) and research in nontraditional areas of mathematics. In this connection, there is a recent report on the recognition and reward system in the mathematical sciences (Joint Policy Board for Mathematics, 1994). Such issues are particularly worrying for junior mathematicians, since it would be unusual for nonmathematicians to be asked for help in a tenure or promotion evaluation; mathematics departments might well be reluctant to rely on outsiders for judgments and decisions viewed as a departmental prerogative. Another related issue is the value attached to work in which an existing body of mathematics is applied to another scientific problem area; even if the impact is great and the work represents a significant scientific advance, it is not "new mathematics," and hence may be accorded little weight in an evaluation of research contributions.

Because of the tendency to preserve and protect departmental boundaries, mathematics departments are ill-equipped to cope with questions that inevitably arise if mathematicians become seriously interested in interdisciplinary problems. For example, if a young mathematician is hired as a

BOX 5.1 American Chemical Society Curriculum Standards for Mathematical Course Work

"Students should emerge from an ACS-approved program in chemistry with:

- A firm foundation in the fundamentals and applications of calculus, including knowledge of differential equations and proficiency with partial derivatives.
- An understanding of the basic principles of linear algebra and practical knowledge of statistics with applications to such problems as experimental design, validation of data, and optimization procedures.
- Experience with computers, including programming, numerical and non-numerical algorithms, simulations, data acquisition, and use of databases for information handling and retrieval."

SOURCE: *Undergraduate Professional Education in Chemistry: Guidelines and Evaluation Procedures*, American Chemical Society, Washington, D.C., 1992, p. 11.

numerical analyst and subsequently becomes interested in chemical statistical mechanics, should this be viewed as a loss or a gain?

For academic chemistry departments, analogous principles of departmental autonomy can affect chemists seeking to work with mathematicians. Because theoretical/computational chemists must often demonstrate the applications of their work to experimental areas of chemistry, fundamental work of a mathematical nature—for example, algorithm development or identification of problem features amenable to mathematical attack—may be undervalued. On balance, chemistry departments have more experience in evaluating multidisciplinary research, soliciting judgments as needed from a variety of scientists both inside and outside the department. A further positive effect on interdisciplinary work is that chemistry departments tend to value research that has a significant impact on thinking, research, and practice in chemistry and other areas.

For both fields, the difficulty of interdisciplinary collaboration is exacerbated by the lack of a well-established network of contacts between mathematicians and chemists. On most university campuses, chemistry and mathematics departments are physically separate, so casual daily contact does not occur. An effort is typically required for faculty to attend all the seminars in their own department, let alone in other departments. This reality aggravates the difficulty not only of initiating a collaboration, but also of developing an appreciation of the other discipline's challenges. Faculty members are not immune to misperceptions and stereotypes: chemists may regard mathematicians as unapproachable or uninterested in chemistry problems; mathematicians may not realize that chemistry problems contain interesting and novel mathematics.

There are, however, exceptions. For instance, in the United Kingdom there is a long fruitful history of productive mathematical research being initiated by theoretical scientists ("natural philosophers") employed as faculty members of mathematics departments. This goes back to Newton, but the tradition continues to modern times. D.R. Hartree and P.A.M. Dirac are recent examples at Cambridge, and C.A. Coulson spent most of his career as a professor of mathematics at Oxford. Some departments in the United States also have established atmospheres that are conducive to collaborative work.

Outside of academia (e.g., pharmaceutical companies), cross-disciplinary work between chemists and mathematicians has succeeded because the problems of disciplinary boundaries are less pervasive in many instances, and team efforts are often the norm. Issues of tenure, grants, and promotion are nonexistent or less important. These settings should provide models for collaborative research.

Effects of the Curriculum

The disparate natures of mathematics and chemistry have led to different training frameworks, which in turn continuously influence and are influenced by the research style and practices in the two fields. The typical curricula encountered by students of chemistry and mathematics, both undergraduate and graduate, do not help to decrease the gaps described above.

In mathematics, basic courses rarely involve exposure to the physical "roots" of problems; mathematicians study idealized problems as exemplars, not for details of the real-world problem. Part of the gap specifically between mathematics and chemistry can be explained by long-standing pedagogical practices in mathematics. Much of classical applied mathematics is based on constructions associated with mechanics and physics: every student of mathematics studies the heat equation, elastic rods, electrical networks, and fluid flow. However, no problems explicitly associated with chemistry are widely taught to or recognized by mathematicians.

There is little time or incentive for mathematics students to learn chemistry at a substantive level, let alone to study interesting chemistry problems. At the undergraduate level, some mathematics curricula require courses in a physical science, but these are more often in physics than chemistry. Although undergraduate mathematicians sometimes take freshman chemistry (frequently a descriptive course), they are unlikely to study physical or organic chemistry. Graduate students in mathematics do not typically take many courses outside their own department and hence have no convenient mechanism for learning about mathematical problems in chemistry.

Education of chemists currently involves little exposure to advanced concepts in modern mathematics (see Box 5.1). Undergraduate chemists take calculus and (perhaps) ordinary differential equations, linear algebra, or numerical methods, but seldom study abstract algebra, differential geometry, numerical analysis, partial differential equations, probability, or topology. Graduate students in chemistry rarely take courses in mathematics departments. Some chemists believe that undergraduate chemistry courses do not require high-level mathematics and prefer instead to build chemical intuition by descriptive methods. For chemistry students and faculty interested in learning modern mathematics, the mathematics curriculum is structured like a tree, with courses of potential interest to chemists at the end of a very long branch of prerequisites; the effect is to discourage chemists from obtaining any knowledge of advanced topics.

Language Differences

Language barriers ranging from conspicuous to subtle must be overcome by anyone who wishes to pursue interdisciplinary work between mathematics and chemistry. At the most obvious level, specialization and the internal communication requirements of mathematics and chemistry have created two technical languages. Thus, fundamental concepts that occur in only one field need to be defined either in the technical language of the other or in a common natural language. Mathematical examples include Pisot numbers, ambient isotropy, and wavelets; chemical examples include ligand, pharmacophore, and racemate.

Within mathematics, each research subdiscipline continually refines concepts and introduces new technical jargon, making it very difficult even for mathematicians in slightly different research areas to communicate with each other. Similar language problems exist in chemistry, though perhaps to a lesser degree. On the mathematical side, the problem is compounded by preferences for an abbreviated writing style. The "abstract minimalism" approach to writing taken in much of the mathematics literature can make graduate texts and research papers in mathematics almost impenetrable except to the most determined readers.

BOX 5.2 Information Sources About Theoretical/Computational Chemistry

For a mathematician who wants to get involved with theoretical and computational chemistry, the best source of information should be the chemists at his or her own institution or research center. The American Chemical Society (ACS) publishes a directory (*Directory of Graduate Research*) that lists the research interests and recent publications of academic chemists, if one wishes to look further afield. Those mathematical scientists who have established productive collaborative or interdisciplinary lines of research often observe that one must be a good listener and be willing to devote time and energy to learning nuances of language and concepts.

There are a number of printed reviews available. For molecular modeling, the book series *Reviews in Computational Chemistry* (D. Boyd and K. Lipkowitz, eds., VCH Publishers, New York), gives a good overview of the field. Also, a visit to the exposition at an ACS national meeting (held twice a year) will give a feeling for the large number of software vendors in this field, the type of software available, and the types of problems of interest to chemists. The November 1993 issue of *Chemical Reviews* was also devoted to this subject. A variety of approaches to computer-based drug design are discussed in the series *Comprehensive Medicinal Chemistry* (Pergamon Press, Oxford).

For electronic structure problems, the literature is very scattered. The book series *Advances in Quantum Chemistry* reviews the more mathematical aspects of the field along with some very applied results. The July/August 1991 issue of *Chemical Reviews* contained reviews on a wide range of applications. There is unfortunately no comprehensive review of the algorithms involved in popular programs, although the user's guides to GAUSSIAN, MELD, and HONDO list many of the papers on which these programs are based. Also, the *Modern Techniques in Computational Chemistry* reports (E. Clementi, ed., ESCOM Science Publishers, The Netherlands) discuss many algorithms. The book series *Relativistic and Electron Correlation Effects in Molecules and Solids* (G.L. Malli, ed., Plenum Press, New York) and the series *Methods in Computational Molecular Physics* (G.H.F. Diercksen and S. Wilson, eds., Reidel Publishing, Dordrecht) also contain several volumes devoted to methods for quantum chemistry. The annual "Sanibel" meeting organized by the Quantum Theory Project at the University of Florida (and now held at St. Augustine, Florida) is a good place to meet quantum chemists. The papers from that meeting are published annually in a special symposium series from the *International Journal of Quantum Chemistry*.

An excellent discussion of the molecular dynamics method, Monte Carlo calculations, and related methods for computer simulation studies of materials is contained in *Computer Simulation of Liquids* (M.P. Allen and D.J. Tildesley, Oxford University Press, 1987). This monograph is a useful resource for learning about the theory of such simulations as well as the algorithms used in research.

Finally, the Computational Chemistry List is a very active electronic clearinghouse for information on that subject. Interested readers may subscribe by sending their name, affiliation, and electronic mail address to chemistry-request@osc.edu.

In addition to separate sets of terminology, mathematicians and chemists face several varieties of linguistic confusion. Similar concepts in both disciplines are sometimes denoted by different words. For example, a (nontrivial) "link" to a mathematician is a collection of elastic circles that are mutually entangled and cannot be separated spatially into subcollections; to a chemist, a "catenane" (from the Latin "catena" or chain) is a collection of circular molecules held together by topological bonds, not by chemical bonds. Mathematics invokes the concept of graph isomorphism type, whereas chemists speak of connectivity of the molecular graph.

At the other extreme, identical technical names may be used by the two disciplines for different concepts or for concepts that are similar but vary in precision. For example, the term "topology" is

BOX 5.3 Information Sources About the Mathematical Sciences

Journals and Newsletters

- *Acta Numerica* contains an annual review article on important developments in numerical analysis.
- *Bulletin of the Institute of Mathematical Statistics* contains information on meetings and articles on items of current interest in statistics.
- *Mathematical Reviews* contains short reviews of articles on a wide spectrum of mathematical topics.
- *Proceedings of the International Congress of Mathematicians* is published every four years and contains invited survey articles by leading mathematicians on recent mathematical progress.
- *Notices* of the American Mathematical Society contains information on meetings and articles on items of current interest in core mathematics.
- *SIAM News* is the news journal of the Society for Industrial and Applied Mathematics, containing information on meetings and items of current interest in applied and applicable mathematics.
- *SIAM Review* contains technical review papers on subjects in applied and numerical mathematics.

Electronic Information

- e-MATH is the electronic information connection to the American Mathematical Society. Information is available on meetings, membership, electronic sources of mathematics information, etc. Telnet e-math.ams.org; login and password are both "e-math" (lower case).
- NA DIGEST is a numerical analysis electronic newsletter; for information, send e-mail to na.help@na-net.ornl.gov.
- NETLIB is a database of public-domain computer programs with e-mail address netlib@ornl.gov.

used in science as a catchall word describing shape phenomena, but it has an exact technical meaning in mathematics. The word "homotopic" can refer to deformation of paths in mathematics and to interconvertible protons in nuclear magnetic resonance (NMR) spectroscopy.

Toward a Fruitful Collaboration

Because research inherently moves into the unknown, there is no way to predict reliably which areas of mathematics and chemistry might work together effectively. Interactions between computational chemists and computational mathematicians are perhaps the most obviously rewarding today for addressing large-scale computational problems that occur in quantum chemistry, molecular mechanics, and molecular dynamics; these areas seem promising for serious collaborations since progress is likely to be made only by combining significant expertise in chemistry, mathematics, and computer science. However, the ranges of opportunity and success stories are very broad (see Chapters 3 and 4 of this report), and fruitful interactions may emerge between chemists and mathematicians in any subfield.

The ideal interdisciplinary collaboration often begins with personal contacts between two or more scientists who share an interest in a particular problem. Just as the experimental chemist might tend to approach the theoretical/computational chemist for assistance in certain areas, it is perhaps more common for a chemist to initiate a mathematics-chemistry collaboration. Recently, however, there has been an increasing emphasis on involvement by mathematical scientists in "grand challenge" problems, and some applied mathematicians have actively sought connections with chemistry.

Once a contact has been initiated, the success of a collaboration depends on a strong sense of mutual respect and benefit among the participants. These feelings are essential so that each partner is willing to learn new science as needed (e.g., protein chemistry for the mathematician, optimization theory for the chemist) and to adapt to a somewhat different style than he or she is accustomed to, welcoming a combination of theory, computation, and physical intuition toward solution of a problem. Such a synergistic process allows an "evolution" of solutions that can progressively address more of the complexity of the realistic problem and incorporate new physical data as they become available.

References

Joint Policy Board for Mathematics, 1994, *Recognition and Rewards in the Mathematical Sciences*, American Mathematical Society, Providence, R.I.

National Research Council and Institute of Medicine, 1990, *Interdisciplinary Research: Promoting Collaboration Between the Life Sciences and Medicine and the Physical Sciences and Engineering*, National Academy Press, Washington, D.C.

6
CONCLUSIONS AND RECOMMENDATIONS

The committee has examined evidence supplied to it in the form of prior reports, expert testimony at its meetings, selected studies of the scientific literature, and personal contacts in the mathematical sciences and chemistry communities. As a result of these investigations and its collective evaluation of the available information, the committee has reached the following conclusions.

- Several notable "success stories" can be identified, illustrating the value of interdisciplinary stimulation and synergistic research collaboration involving cooperation between the mathematical sciences and the theoretical/computational chemistry communities.
- Many opportunities appear to exist for further collaborations between the mathematical and chemical sciences that could result in high-quality scholarship and research progress that would advance national interests. The productivity of applied computational chemistry would likely be enhanced as a result, which could be potentially significant for industry.
- Active encouragement of further collaborations is warranted because it would likely result in an acceleration of such research progress.
- Cultural differences between the mathematics and the chemistry communities, involving language, training, aesthetics, and research style, have tended to act as barriers to collaboration, even in circumstances that might otherwise suggest the benefit of cooperation.
- Institutional structures and reward systems in the academic community have often placed significant difficulties in the way of collaborative research across traditional disciplinary boundaries, which can be especially inhibiting to those in early career stages.
- Government funding agencies have for the most part made constructive efforts to identify and fund worthy interdisciplinary and collaborative research. However, this process is still somewhat haphazard. Agencies tend to be organized along traditional disciplinary lines, and the evaluation of interdisciplinary proposals relies on personal contacts between program managers and on timely and comprehensive responses from what is typically a small pool of qualified reviewers. The time lapse involved in the proposal evaluation process thus has often been anomalously long.
- To a large extent, both mathematical scientists and theoretical/computational chemists are relatively unaware of the most exciting recent advances in each others' fields. Consequently both groups tend to be insensitive to the opportunities for interdisciplinary cross-fertilization that could produce intellectual novelty and productivity enhancements on both sides.
- The system of prizes and awards administered by the mathematical sciences and chemistry professional societies is currently not geared to recognize and reward interdisciplinary collaborative research advances.
- The national environment—including Congress, funding agencies, and the professional societies (see, e.g., Joint Policy Board for Mathematics, 1994)—has become perceptibly more conducive to encouraging and supporting interdisciplinary and collaborative research, particularly as it may concern industrial innovation and productivity. Government agencies in particular are currently in a mood to actively encourage joint industrial-academic research, even though proprietary rights barriers to free collaboration are recognized to exist.
- The overwhelming volume of specialized technical literature aggravates the communication problems between fields and occasionally leads to wasted effort, redundancy, and rediscovery. It appears that well-researched and well-written review articles spanning normally disconnected specialties in the mathematical sciences and in theoretical/computational chemistry represent a disproportionately small fraction of the technical literature, in spite of the fact that they can eliminate redundant effort.

In response to these conclusions and to the insights gained from its study, the committee makes the following recommendations:

Undergraduate Education. The best way to attract scientists to interdisciplinary work is to get them interested as undergraduates. It is recommended that universities encourage undergraduate interdisciplinary research courses, seminars, and summer programs. For example, mathematical sciences departments could institute seminars for undergraduates in which chemists (and other scientists) would be invited to discuss chemistry research areas that might benefit from interaction with mathematics. The committee recommends that chemistry departments establish seminars for undergraduates in which mathematical scientists would be invited to discuss modern mathematics. Graduate students (and interested faculty) would of course be welcome to attend these seminars.

In the experience of the committee members, one very successful vehicle for getting mathematics and chemistry undergraduates interested in research is the REU (Research Experience for Undergraduates) program sponsored by the National Science Foundation (NSF). In addition to fostering interdisciplinary undergraduate activity at research universities, there is a real educational opportunity here for four-year liberal arts institutions that traditionally encourage undergraduates to write senior honors theses and to otherwise construct, expand, and explore their own undergraduate education.

Graduate Education. Departments in the mathematical and chemical sciences should encourage graduate degrees (both M.S. and Ph.D.) that involve dual (mathematics and chemistry) mentoring. Dual mentoring activity between chemistry and physics and chemistry and biology has been successful in many universities. The committee recommends that mathematics graduate students consider a minor in chemistry instead of a minor in an area of mathematics related to their research specialty. Theoretical and computational chemistry graduate students should consider a minor in mathematics or, alternatively, take a core of mathematical courses appropriate to their interest (perhaps in the framework of a special "interdisciplinary track"). One way to encourage cross-disciplinary graduate education is to allow graduate students in one area to enroll in upper-level undergraduate courses in another area for graduate credit.

Faculty Interaction. Mathematics and chemistry departments should on occasion invite a person from the other area to speak in a research seminar or a colloquium. Lists of speakers of potential interest to industry should be circulated to local industrial laboratories, and vice versa.

Interdisciplinary Research. The committee recommends that mathematics and chemistry departments encourage and value individual and collaborative research that is at the interface of the two disciplines. Such work has the potential for significant intellectual impact on computational chemistry, and hence on the future evolution of chemical research and its applications to problems of importance in our society.

Professional Societies. The American Mathematical Society (AMS) issued a policy statement in 1994 that supports interdisciplinary research. The second goal of that statement is to "connect the power of mathematics and mathematical thinking to problems in science, technology, and society." This policy statement is reinforced by specific recommendations to "enlarge the scope and extent of interdisciplinary research connecting mathematics with other fields" and to "emphasize the value of such connections during the mathematical training of both undergraduate and graduate students" (American Mathematical Society, undated).

Professional meetings in mathematics and chemistry—for instance, those of the AMS, American

Chemical Society, Society for Industrial and Applied Mathematics (SIAM), and the Chemical Physics Division of the American Physical Society—would benefit from talks very much like the seminar and colloquium talks described in the recommendation for faculty interaction above, from shorter presentations in special sessions, and from panel discussions. There are already some promising moves in this direction as reflected, for example, by recent AMS sessions on mathematics and molecular biology or SIAM sessions on molecular chemistry problems and optimization. These sessions at national and regional professional society meetings could generate a mailing list of interested mathematicians and chemists, which could be the kernel of a community of people interested in interdisciplinary research. Appropriate specialized interest groups such as already exist for other fields might be established. One way to encourage these interests is by initiating small interdisciplinary workshops, perhaps incorporating tutorials for students. Examples of successful interdisciplinary meetings are the mathematics and molecular biology series in Santa Fe and the mathematical physiology series held at the Mathematical Sciences Research Institute at the University of California at Berkeley. Funding agencies have in the past funded carefully planned interdisciplinary meetings.

Prizes and Awards. The committee recommends that professional societies in the mathematical and chemical sciences examine the feasibility of establishing awards and named lectureships for work at the mathematics-chemistry interface. High-level public recognition by peers would be a major step toward breaking down interdisciplinary barriers.

Expository Articles and Books. Professional journals in mathematics and chemistry could enhance their quality, appeal, and influence by publishing expository articles on work at the mathematics-chemistry interface. There is a shortage of books written for someone who is mathematically (chemically) sophisticated and desires fairly precise but nonrigorous chemical (mathematical) explanations. The committee encourages mathematicians and chemists to write expository books aimed at this interdisciplinary area.

Interdisciplinary and Industrial Postdoctorals and Sabbaticals. Mathematics and chemistry departments should encourage postdoctoral and faculty sabbatical study at the mathematics-chemistry interface. The committee recommends that the chemical software, pharmaceutical, and chemical industries expand their use of mathematics postdoctorals and faculty on sabbatical leave, and increase their cooperation with and utilization of existing NSF programs such as the University-Industry Cooperative Research Program in the Mathematical Sciences; Industry-Based Graduate Research Assistantships and Cooperative Fellowships in the Mathematical Sciences; Mathematical Sciences University-Industry Postdoctoral Research Fellowships; and Mathematical Sciences University-Industry Senior Research Fellowships. Another opportunity in this regard exists at the Institute for Mathematics and Its Applications at the University of Minnesota, which has an active industrial postdoctoral research program with the aim of broadening the perspectives of recent doctoral recipients in the mathematical sciences and preparing them for research careers involving industrial interaction.

References

American Mathematical Society, undated, AMS National Policy Statement 94–95: Summary, American Mathematical Society, Providence, R.I.

Joint Policy Board for Mathematics, 1994, *Recognition and Rewards in the Mathematical Sciences*, American Mathematical Society, Providence, R.I.

AFTERWORD

In the course of its study the committee returned several times to the question of how to maximize the impact of its work and how to measure that impact. To achieve maximum influence, leaders of the mathematical sciences community and researchers at the interface of the mathematical sciences and chemistry were invited to supply input to the committee to ensure that a broad range of experience and expertise was sampled. Emphasis was put on writing a report that would appeal both to mathematical scientists and to theoretical/computational chemists. A dissemination effort including electronic and hard-copy publication was planned to make the report widely available and to convey its recommendations to a variety of community leaders and policymakers.

Because the promise of interdisciplinary work between the mathematical sciences and theoretical/computational chemistry is so great, the committee suggests that the cognizant boards of the National Research Council (NRC) invite testimony five years hence from community leaders and federal program managers to assess whether progress has been made in achieving this promise. It would be worth attempting to gauge specifically whether the present report helped to bridge the stylistic and linguistic gap between the two fields addressed, especially since future NRC studies on interdisciplinary topics might benefit from the committee's experiences.

GLOSSARY

The following entries represent words, phrases, abbreviations, and acronyms that appear in this report. In many cases, definitions have been given in the text itself, but not with each occurrence. This collection is not offered as an exhaustive mathematics-chemistry bilingual dictionary, but rather as a representative guide to the kinds of technical terms often used by the respective communities and an aid to the reader. It contains examples of multiple meanings that can degrade communication between well-intentioned, but differently trained, professionals. Even without reading the main text, a perusal of this Glossary may convey a sense of the linguistic barriers that occasionally inhibit effective collaboration.

Ab initio method Usually, quantum chemical computation procedures that explicitly include all electrons and utilize their full Hamiltonian operator; occasionally used to describe pseudopotential methods or other approaches with minimal semiempirical input.

ACS American Chemical Society.

Adiabatic Sufficiently slow variation of the externally controllable parameters of a quantum mechanical system so that the quantum eigenstate occupation probabilities remain unchanged over time.

Adjacency matrix For an n-atom molecule, an $n \times n$ matrix with unity as the i, jth entry if atoms i and j share a covalent bond; otherwise it is zero.

Advection Similar in spirit to "convection," but referring to forced flow under more general circumstances.

Affinity Capacity for binding between two molecular units that leads to a more or less stable chemical combination.

Alkaloids Nitrogen-containing organic compounds of natural origin that act as bases in solution; examples are nicotine, quinine, morphine, and lysergic acid.

Alkane Saturated hydrocarbon (i.e., containing only single bonds between pairs of carbon atoms and carbon-hydrogen pairs). In the strict usage this refers to molecules in which the skeleton of carbon-carbon single bonds has the form of a Cayley tree and so exhibits the chemical formula C_nH_{2n+2}; occasionally it refers to saturated hydrocarbons with cyclic skeletal closures (cycloalkanes) that have lower relative hydrogen content.

Amino acid Organic acid containing an amino (nitrogen-hydrogen) group and a carboxyl group, usually in neighboring locations; the modular building blocks of polypeptides and proteins.

AMS American Mathematical Society.

Anionic site Location in a protein, gel, catalytic solid, etc., that bears an excess negative charge of one or more fundamental units (the electron charge).

APS American Physical Society.

Assay Experimental procedure to determine the magnitude of a property of a substance. Usually used in a biological or biochemical context.

Band reduction Algorithmic process applied to matrices to result in all nonzero terms being close to the diagonal.

Bar-coding Familiar supermarket technology applied to efficient automated screening of compounds in the pharmaceutical industry: physical labels for packages with patterns of black and white parallel bars that encode text and numeric information.

Bechgaard salts Organoselenium ionic compounds exhibiting low-temperature superconductivity.

Biological assay An experimental procedure to test molecules for their ability to elicit a biochemical or pharmacological response.

Born-Oppenheimer approximation, potential surface Approximate separation of variables in atomic and molecular quantum mechanics justified by the large ratio of nuclear to electron masses; also the potential energy surface resulting therefrom, as a function of nuclear positions.

Breit-Pauli relativistic corrections An approximate correction for relativistic effects, used as a perturbation to solutions of the intrinsically nonrelativistic Schrödinger wave equation.

Buckyball Nickname for fullerene (see below). Usually applied to the prototypical C_{60} molecule.

Catalytic antibodies Proteins central to the immune response that have been artificially endowed with catalytic properties.

Catenane Chemical compound consisting of two or more ring molecules that are unconnected by chemical bonds but linked by topological entanglement.

Cayley tree Linear graph (vertices linked by bonds) that is connected but contains no closed polygons.

Chirality Geometric or topological property of a molecule or other structured object distinguishing it from its mirror image (i.e., "handedness").

Clique detection Computational procedure to identify the common subgraph(s) of a pair of graphs, especially the maximal common subgraph.

Cluster analysis, clustering The grouping of chemical compounds, usually on the basis of many distinct attributes, so that members of any one grouping resemble one another, but are clearly different from those of all other groupings.

COLUMBUS Software package for quantum chemical calculations.

Combinatorial synthesis (combinatorial chemistry) A procedure for the simultaneous synthesis of a collection of related molecules, usually accomplished on a polymeric support.

Condensed phase A state of matter in which the constituent particles (atoms or molecules) are densely packed in space and in constant interaction with neighbors; usually refers to liquids, liquid crystals, glasses, crystalline solids, or quasicrystals.

Conformation, conformation space Geometric specification of the spatial arrangement of a molecule, possibly flexible; the space of all possible arrangements.

Cut-and-projection method Technique for generating quasi-crystal structures by isolating slabs from higher-dimensional periodic structures and projecting them into three-dimensional space.

DARC Commercial software for managing chemical information databases.

Daylight Chemical database software.

Density matrix Hermitian matrix composed as the direct product of a wavefunction and its complex conjugate. "Reduced density matrices" are obtained by integrating over some of the variables in the full density matrix.

Distance geometry Reconstruction of the full three-dimensional shape of a protein or other biopolymer from a given set of intramolecular distances, which in some applications are known only between limits.

DNA plasmid An autonomously replicating circular DNA molecule.

Docking The fitting and binding of small molecules (ligands) at the active sites of biomacromolecules (e.g., enzymes and DNA).

Dopamine receptor The protein on the cell surface that recognizes the neurotransmitter dopamine and turns on cell responses as a consequence of this recognition.

Electrophoresis An experimental technique to fractionate (separate) charged molecules that relies on differential migration velocities through an obstructive medium (gel), under the influence of an applied electric field.

Enantiomer Mirror image counterpart of a chiral molecule.

Ergodicity Capacity of a dynamical system spontaneously to sample all of its phase space.

Euler's gamma function Analytic extension of the factorial function from the positive integers to the complex plane.

EVB An approach to an approximate solution of the electronic Schrödinger equation, using an empirical valence bond wavefunction.

FCC crystal Face-centered cubic crystal, one of the close-packed arrangements of rigid spheres in three-dimensional space.

Fermi resonance Coupling between normal modes of vibration due to anharmonic potential energy contributions, when a resonance condition is satisfied (rational ratio of normal mode frequencies).

Fermion Particle with half-integer spin, and therefore exhibiting wavefunction antisymmetry and Fermi statistics.

FFT Fast Fourier transform.

Fluxional molecule Flexible molecule, capable of substantially changing its conformation (shape) under the prevailing temperature and surrounding-medium conditions.

Fullerene Stable molecule consisting entirely of carbon atoms arranged at the vertices of a convex polyhedron.

GAMESS Software package for quantum chemical calculations.

GAUSSIAN Software package for quantum chemical calculations that emphasizes use of Gaussian basis functions.

Hamiltonian In classical mechanics, the sum of kinetic and potential energy functions (i.e., the total energy); in quantum mechanics, the corresponding linear Hermitian operator.

Harmonic analysis A mathematical theory that aims to decompose complex phenomena into the superposition of simpler phenomena (such as wave forms).

Hartree-Fock approximation Replacement of an exact wavefunction with an antisymmetrized product of single-particle orbitals (i.e., a Slater determinant).

Hermitian operator Linear operator whose matrix elements exhibit the property that reflection across the diagonal is equivalent to complex conjugation.

Hessian matrix Symmetric matrix of second partial derivatives, often used in optimization routines for many types of objective functions. A frequently encountered application involves searching for extremes on potential energy surfaces.

Heteropolymer A polymer formed from nonequivalent monomer units.

High-performance computers A phrase adopted to refer to emerging computers of a variety of designs (parallel, vector, and others) that reflects the convergence of many of these design concepts.

Homotopic paths Paths that share the same end points and can be continuously deformed into one another while those end points are kept fixed.

HONDO Software package for quantum chemical calculations.

Hückel theory An approximate theory of molecular electronic structure that uses a minimal basis of atomic orbitals, and simplified Coulomb and exchange integrals.

Human genome (project) The 3×10^9 base pairs that constitute the entire human genetic heritage; the Human Genome Project aims for a complete mapping of the human genome, identifying and sequencing regions of chromosomes that code for individual proteins.

Hydrogen bond Weak noncovalent interaction (mainly electrostatic) bond between a significantly positive hydrogen atom and an electronegative atom (usually nitrogen, oxygen, or fluorine); consequently, an important structural element in water, proteins, and other biopolymers, as well as in the recognition (specific binding) of small molecules by biopolymers.

Interval analysis A mathematical technique that involves computation of strict bounds to bracket the global minimum of a function.

Irreducible representation Group representation (e.g., as matrices) that does not admit of decomposition into a direct product of elements of lower dimension.

JPBM Joint Policy Board for Mathematics.

Koopmans's "theorem" Asserts that orbital eigenvalues of the Hartree-Fock approximation provide good estimates of the vertical (fixed nucleus) ionization energies of atoms or molecules.

Langevin equation, dynamics An approach for analysis and simulation of molecular motion in which the molecular forces in Newton's classical equations are modified by additional terms (a frictional kernel and stochastic forces) that attempt to describe in a simple and computationally feasible way the effects of surroundings (e.g., solvent medium) on motion of the molecule of interest.

Lie group A differentiable manifold that has a group structure on its elements, with the property that the group operations (multiplication and inversion) are continuous. A standard and useful example is the set of nonsingular real $n \times n$ matrices. Other examples are the group of rigid motions in Euclidean space and the group of equivalence classes of transformations that agree on some neighborhood of the identity. The last was the original object of study by Sophus Lie around 1890.

Ligand A molecule or other chemical grouping attached to a larger molecular structure. In the case of a small molecule that binds to a biomacromolecule, the latter is frequently referred to as the "receptor."

MAA Mathematical Association of America.

MDI Commercial software for chemical information databases (Molecular Design Incorporated, San Leandro, California).

MELD Software package for quantum chemical calculations.

Molecular dynamics (MD) Computer simulation technique for many-body systems that relies on numerical solution of classical equations of motion for atoms or molecules and evaluates thermodynamic, kinetic, and structural properties as time averages.

Molecular mechanics The field encompassing molecular statics, or the construction of appropriate force fields for representing molecular systems and the associated potential energy minimizations.

Molecular modeling The art of representing and analyzing molecular systems by using mathematical concepts and techniques, numerical computation, graphics, etc.

Monte Carlo methods Numerical evaluation procedures relying on probability and the repeated use of random number generators.

NMR Nuclear magnetic resonance.

Nodal cells Connected regions of the configuration space for a set of electrons, over which their wavefunction maintains constancy of sign.

NOE Nuclear Overhauser effect.

Normal modes Independent harmonic motions of a mechanical system that possesses a quadratic potential energy function; alternatively, such motions of small amplitude on a more complicated surface that can be locally approximated as quadratic.

***N*-representability** Attribute of candidates for reduced density matrices (in quantum mechanics) or for molecular distribution functions (in classical statistical mechanics) that they correspond to the contraction of some legitimate N-body wavefunction or distribution function.

NSF National Science Foundation.

Nucleotide Hydrolysis product (monomeric unit) of nucleic acids.

***n*-valent vertices** Vertex of a linear graph at which n bonds (edges) are attached.

Padé approximant Rational function approximation to an analytic function, usually determined by matching power series coefficients.

Parallel computer A computer with multiple processors working in parallel on portions of one computation.

Path integral An appropriately weighted sum over a family of line integrals connecting fixed end points. Applications include descriptions of Brownian motion and quantum transition probabilities.

Pauli exclusion principle The requirement for electrons (or other fermions) that the wavefunction for a collection of such particles is antisymmetric under interchange of the position and spin coordinates of any pair of identical particles. In the case of noninteracting or uncorrelated particles, it leads to the requirement that at most, one particle can be in each single-particle state.

Pharmacophore The chemical identity and geometrical arrangement of key substituents in a molecule that confer biochemical or pharmacological effects.

Phase In quantum mechanics, the phase angle of a complex wavefunction or order parameter; in statistical mechanics, the state of aggregation of matter (e.g., crystal, liquid, vapor, quasicrystal, liquid crystal); in dynamics, the position-momentum specification.

Phase space The joint space of configurational coordinates and their conjugate momenta for a classical dynamical system.

Pisot numbers Root of a polynomial with integer coefficients, all of whose other roots have absolute values less than unity.

Polyacetylene Linear polymer bearing delocalized electrons, formed from acetylene (C_2H_2).

Polyvinylcarbazole Photoconductive polymer that has been used as an active material in xerographic applications.

Potential energy surface Surface (hypersurface) in ($3N + 1$)-dimensional space expressing the potential energy of interaction within an N-particle system, as a function of the $3N$ position coordinates; see Born-Oppenheimer approximation.

Prokaryote A single-cell organism that has no distinct nucleus.

Protein folding Spontaneous development, under ambient conditions, of the natural three-dimensional shape of a protein molecule that facilitates its biological function.

Pseudorandom number generator Numerical algorithm to approximate an ideal generator of statistically independent, uniformly distributed, random numbers.

QSAR Quantitative structure-activity relationship. A statistical method that relates biological potency to physical properties of molecules.

Quantum chaos Dynamical evolution of a quantum mechanical system that displays irregular motions and extreme sensitivity to initial conditions.

Quasicrystals Solids exhibiting long-range orientational coherence of local atomic coordination geometry but no crystallographic periodicity or unit cell.

Quasi-ergodic Property of a dynamical system that it eventually visits a well-defined neighborhood of any kinematically accessible point in its phase space.

Racemates Mixtures containing equal amounts of enantiomeric (mirror-image stereoisomer) molecules.

Raman spectroscopy Technique of molecular spectroscopy enabled by the coupling of vibrations to electronic polarizability.

Random number generator A random number generator is a computer procedure that scrambles the bits of a current number or set of numbers in such a way that the result appears to be randomly distributed among the set of possible numbers and to be independent of the previously generated numbers. The lagged-Fibonacci random number generator, for instance, uses a shift operation and a binary operation on n-tuples from a finite set (usually the integers mod m).

Reaction coordinate Configurational coordinate measuring distance across a transition state saddle point on the potential surface of a chemically reactive system.

Residue A chemical unit, such as an amino acid, of a large molecule.

REU The Research Experience for Undergraduates program of the National Science Foundation.

Ribosome Cellular structure that is the usual site of protein synthesis in living organisms.

Riemann zeta function Sum of the inverse *s*th powers of the positive integers, viewed as an analytic function of *s*.

Salem numbers Root of a polynomial with integer coefficients, all of whose other roots have absolute values equal to or less than unity.

Scalable parallel algorithms or parallel computers Those that can continue to perform efficiently as both the problem size and the number of processors increase.

Schur's lemma Relation between irreducible representations of a finite group.

Semiclassical approximation Treatment of dynamics in which quantum effects are regarded as a weak perturbation on classical mechanics; often implemented formally as the limit at which Planck's constant approaches zero.

Sequence space The set of possible ordered monomer sequences for a biopolymer. For proteins this is the "primary structure," the amino acid sequence along the molecular backbone; for DNA it is the sequence gene-encoding bases (adenine, cytosine, guanine, thymine).

Sequential computers (uniprocessors) Computers that operate on individual pieces of data in sequence. Contrast with vector processing and parallel computers.

SIAM Society for Industrial and Applied Mathematics.

Slater determinant Determinant whose elements are *n* distinct orbitals for *n* distinct electrons.

Solitons Stable pulse or particle-like solutions of some nonlinear wave equations.

Spin glass A large collection of spins whose interactions frustrate simple ordering and thus create many deep potential energy minima that are nearly degenerate.

Statistical mechanics The study of the collective behavior of large numbers of interacting particles. Properties of interest include those describing time-dependent, irreversible process. The basic principles of this discipline were laid down in the nineteenth century by Ludwig Boltzmann, James Clerk Maxwell, and Josiah Willard Gibbs.

Stereoisomers Molecules that differ only by mirror inversion of bonds at chiral centers (usually carbon atoms).

Stochastic dynamics A framework for expressing the dynamics of a molecular system that includes stochastic elements (e.g., random forces that mimic effects of the environment); see **Molecular dynamics** and **Langevin equation**.

Supercoiling The interwinding of double helical DNA upon itself; also called supertwisting.

Symplectic integrator Class of numerical algorithms for integrating classical many-body equations of motion, that exactly preserve phase space volume and other classical invariants of motion.

Terpenes Unsaturated (double-bond-containing) hydrocarbons with composition $C_{10}H_{16}$; typically aromatic oils of natural origin.

Topological (geometric) phase Phase factor appearing in a nuclear wavefunction that arises from, and varies continuously around, a conical intersection of two electronic potential surfaces.

Transition state Saddle point on a potential energy surface separating chemical reactants from products.

Tridiagonalization Transformation of a symmetric matrix so that the only nonzero elements are those along the principal diagonal and its two immediately flanking diagonals.

Ultrametric, ultrametricity Distance (metric) structure applied to a set of objects arrayed on a Cayley tree; for a given pair of objects it is the minimum number of steps along branches of the tree necessary to effect connection.

Unrestricted Hartree-Fock approximation Use of an approximate wavefunction consisting of a sum of Slater determinants, wherein distinct spatial orbitals are employed for distinct spin directions.

Upwinding schemes Techniques for solving advection problems that take into account force direction to produce more accurate results.

Vectorization Adaptation of a computer program to take advantage of vector processing capacity.

Vector processing A mode of computing (using "vector computers") that relies heavily on the concurrent processing of the elements of (possibly large) linear arrays.

***V*-representability** Special form of *N*-representability (see above) guaranteeing that the density matrix or molecular distribution function of interest corresponds to an *N*-body system with additive pair interactions (specifically Coulomb interactions in the atomic or molecular quantum cases).

Wavelets Basis functions selected to represent efficiently certain types of frequently encountered pattern elements, such as discontinuities.

Yield Fraction (normally expressed as a percentage) of the chemical reactants that proceed to the desired product substance.

Zeolite Solid silicates with open crystal structures permitting action as ion exchange media, catalysts, or molecular sieves.